A series of student texts in

CONTEMPORARY BIOLOGY

General Editors :

Professor E. J. W. Barrington, F.R.S.
Professor Arthur J. Willis
Professor Michael A. Sleigh

An Introduction to Evolutionary Genetics

David T. Parkin
Department of Genetics
University of Nottingham

University Park Press
Baltimore

First published in Great Britain by
Edward Arnold (Publishers) Limited, London

First published in the United States of America by
University Park Press, 233 East Redwood St., Baltimore,
Maryland 21202

Library of Congress Cataloging in Publication Data

Parkin, David T.

An introduction to evolutionary genetics.

(A Series of student texts in contemporary biology)

Bibliography: p.

Includes index.

1. Genetics 2. Evolution I. Title [DNLM:
1. Evolution 2. Genetics QH430 P247i]
QH430. P367 1979 575 79-4866

ISBN 0-8391-1443-5

Printed in Great Britain

Preface

The theory of evolution by natural selection was proposed by Darwin and Wallace in the middle of the 19th Century, long before the rediscovery of Mendel's Laws. This book attempts to describe how the theory has itself evolved and changed in the century from its inception to the present day.

Contemporary evolutionary genetics is a complex science that combines a sophisticated theoretical framework with both a strong laboratory component and detailed field observations. I have endeavoured to review the subject at a fairly elementary level, drawing information from all three branches to give a balanced view of the basic principles. The book has evolved from courses that I have taught at Nottingham over the last seven years and reflects my own biases and interests, as well as those of the undergraduates that I have taught.

It makes no attempt to cover fields that are already represented by books in this series. Thus, the reader will find little or nothing on basic genetics, chromosome structure and evolution, or the evolutionary aspects of developmental biology.

During the time that I have been preparing this book, I have received a great deal of help from many people. In particular, I thank Professor Bryan Clarke, who read most of the manuscript, and Professor Ernest Barrington, who was noble enough to read it all. Drs. Tom Day and Jeremy Greenwood, and my graduate students Eric Verspoor and David Whitehouse, helped more than they realise by discussion and criticism of many of the topics. In addition, the undergraduates helped by their interest and disinterest, understanding and confusion. Nothing clarifies one's ideas better than having to explain them to others. It is also a pleasure to express my thanks to Petra Gendle who typed the various drafts, and skilfully produced an ordered manuscript from my almost illegible chaos. And finally, to my wife, whose constant encouragement helped me more than I can say.

ACKNOWLEDGEMENTS

I am grateful to the authors, editors and publishers concerned for permission to reproduce some figures that have previously been published elsewhere. These include: 1.1 from *Darwin's Finches*, by D. Lack (Cambridge University Press, Cambridge); 2.6 from *The Evolution of Melanism*, by H. B. D. Kettlewell (Oxford University Press, Oxford); 2.8, 6.2, 7.6 from *Genetics*; 3.4 from *Atlas of Protein Sequences and Structure*, ed. M. D. Dayhoff (National Biomedical Research Council, Washington); 3.6 from *Nature*; 4.4 from *Genetical Variation in Human Populations*, ed. G. A. Harrison (Pergamon, Oxford); 5.3 from *Annals of Eugenics*; 5.5, 5.6 from *Cold Spring Harbor Symposium in Quantitative Biology*; 5.7, 6.3, 6.6, 7.1, 8.10, 8.11 from *Heredity*; 6.1, 8.8 from *Ecological Genetics*, by E. B. Ford (Chapman and Hall); 6.4, 6.5 from *Evolution*; 7.2 from *Scientific American*; 7.7, 7.8 from *Philosophical Transactions of the Royal Society of London*; 7.9 from *Evolution and Environment*, ed. E. T. Drake (Yale University Press); 8.1 from *Vertebrate Palaeontology*, by A. S. Romer (University of Chicago Press); 8.2 from *The Species Problem*, ed. E. Mayr (American Association for the Advancement of Science); 8.4 from *Animal Species and Evolution*, by E. Mayr (Harvard University Press, Cambridge, Mass.); 8.5 from *The Bird Faunas of Africa and its Islands*, by R. E. Moreau (Academic Press, London); and 8.6 from *Ecological Isolation in Birds*, by D. Lack (Blackwell, Oxford).

Nottingham, D.T.P.
1978.

Contents

I

Evolution by Natural Selection

DARWINISM

It seems almost traditional to begin a book like this with the statement that the theory of evolution by natural selection is the greatest unifying theory in biology. Nevertheless, it remains true that this one concept draws together a vast range of otherwise separate studies of the variation and distribution of living and extinct organisms. Systematics and taxonomy are only given real meaning and direction when coupled with the idea that species evolve and diverge. Similarly, comparative anatomy, physiology and embryology provide little other than academic exercises unless the interrelationships of the phenomena are made apparent by an understanding of the evolutionary process.

The importance of this theory to modern biology makes it necessary to delve at once into the past, to discover why and how Charles Darwin[64] appreciated that evolutionary change takes place through the action of natural selection imposed by the environment. For Darwin did not invent evolution. His greatest contribution to science was to suggest, and demonstrate to the best of his ability, a method by which it could, and, as we now believe, probably did, take place.

In the middle of the last century, the collecting of biological specimens from all round the world had revealed that a vast number of species of animals and plants exist upon the earth. Biologists had described and catalogued them. It had become evident from their studies that species could be arranged into groups on the basis of similarities and differences in their anatomy. Linnaeus had used this as the basis for his nomenclature, and people were now beginning to think that it reflected true relatedness. But why should this be so, and how could such systems of relationship have arisen?

Studies in geology were in the same early descriptive stage as biology. Nevertheless, they were revealing that species might not be permanent entities. There was evidence that forms present in the fossil record were lacking from contemporary faunas. Darwin had observed this for himself when he studied the fossilized remains of giant armadillo-like creatures in South America, which were quite unknown to the field biologists of the day. Conversely, it was beginning to appear likely that some species present in the fauna of the nineteenth century were absent from the fossil remains being uncovered all round the world. The appearance and disappearance of fossil forms, the geological evidence of extinction, likewise required an explanation.

The key was furnished when Darwin's voyage around the world stopped briefly in the Galapagos Archipelago,[65] five hundred miles into the Pacific Ocean off the coast of Ecuador. Here lay a mass of islands varying in size, close to the equator and, sitting in the cold waters of the Humboldt current, well isolated from continental South America. Darwin spent some time among these islands, observing and collecting the curious finches which now bear his name, and also the remarkable giant tortoises.

These two groups of animals were very interesting to him. The finches were clearly similar to one another in many of their morphological characters. For birds of their size, the wings and tail were short, and the feet and legs were rather large. Most of the forms had a plumage which was very similar in colour and pattern. Furthermore, the flight and general behaviour of the different forms had many resemblances. Despite these similarities, the variation in the size and shape of their beaks, and the differences in body size, were sufficient to produce a variety of forms of finch which appeared to be quite distinct from one another. These forms could be regarded as separate species, each being well adapted to its particular role in the Galapagos fauna.

Figure 1.1 shows examples of the various finches to be found on the single island of Santa Cruz. There are three species of seed eaters, with bills of different sizes, which spend much of their time feeding upon the ground.[165] There are arboreal seed eaters with more pointed bills, and species with shorter softer bills which feed upon buds or insects in the trees. One form is so extreme as to be almost identical with the old world leaf warbler (*Phylloscopus*) in size and shape, even flicking its wings in the same nervous manner. Perhaps most remarkable of all is the woodpecker finch which spends much of its time on

Fig. 1.1 The species of Darwin's Finches found upon the Island of Santa Cruz in the Galapagos Archipelago.
Top line The three species of ground feeding seed-eaters. Separated from one another by the size of their bills and the size of the seeds upon which they feed.
Second line *G. difficilis* is now almost certainly extinct. It lived in the humid zone and probably fed upon seeds in the trees. Separated by habitat zone from *G. scandens* which is also an arboreal seed-eater, but from the transitional and arid zones.
Third line *Platyspiza crassirostris* is distinct from all the rest since it feeds upon buds and leaves. *C. psittacula* and *C. parvulus* are both arboreal insectivores separated from one another by bill size and consequently, prey species.
Bottom line Three highly specialized forms. *Cactospiza pallidus*, the woodpecker finch, eats insects from the trunks and branches of trees. *C. heliobates* is exclusively found in the mangrove swamps. *Certhidea olivacea*, the warbler finch, feeds upon small insects collected from the foliage of trees and shrubs.
Thus all these species are distinguished on the combination of habitat and food resources. From Lack.[165]

the branches and tree trunks, picking insects out of the cracks and crevices in the bark with its rather long bill. This species will utilize a cactus spine as a tool to help extraction when the prey is too deep to be reached in a more conventional manner.

Bearing in mind the many anatomical resemblances such as wings, tails and feet, in addition to the internal features and feather number and distribution which are less visible to casual inspection, it was apparent to Darwin that these birds formed a homogeneous group. It seemed possible to him that they had descended from a common ancestor. The differences in anatomy between the species could be explained as adaptation to the particular niches in the environment which they occupied.

He found the tortoises equally remarkable. Giant tortoises occur nowhere in South America apart from the Galapagos Archipelago. Each island population is a distinct variety, and there were even differences between the populations living on the various volcanoes on one single island. In fact, Darwin was very impressed to learn from the governor of the Archipelago that it was possible to identify quite conclusively the island from which a particular tortoise originated. Again, the island forms were clearly closely related to one another, although slight differences in anatomy were to be seen which might or might not be important adaptations to survival.

A third observation which he made during his studies of the Galapagos Islands concerned variations in the flora. Some of his results are summarised in Table 1.1, which lists the number of plant species that he knew to occur on four of the large islands. The species are categorized either as Galapagos natives, or those known to occur

Table 1.1 The number of species of plants reported from four islands of the Galapagos Archipelago by Charles Darwin.

Island	Total number of species	No. of species found in other parts of the world	No. of species confined to the Galapagos Archipelago	No. confined to the one island	No. confined to Galapagos Islands but found on more than one island
James	71	33	38	30	8
Albemarle	46	18	26	22	4
Chatham	32	16	16	12	4
Charles	68	39*	29	21	8

* Or 29 if those plants which have probably been imported into the island by man are excluded.

in other parts of the world. Excluding ten species from Charles Island that he considered to have been introduced by man, Darwin estimated that over half occurred nowhere else in the world and were thus endemic to the archipelago. Furthermore, the majority of these species were in fact endemic to one single island within the group.

Present-day systematics has resulted in great modification and revision of the interrelationships of these plants, but the fact remains that the islands of the Galapagos exhibit an unusually high degree of endemism. Again, there are close similarities between the plants of several islands, and yet they show differences between the islands that are sometimes substantial.

The situation in Galapagos convinced Darwin that evolution is a major process in biology. He appreciated clearly that when a population occupied a new niche it frequently evolved in a different direction to other similar populations. Thus, the tortoises and the finches had reached the islands at some time in the past, and had evolved in a variety of directions to give the various forms present today. Many of the finch species are now found on several islands, but the tortoises and many species of plants only occur on the island where they must have originated.

Darwin's particular genius was that he was able to draw together his field observations with pertinent evidence from a variety of other sources, and to solve the conundrum of evolutionary change. There are three pieces of crucial evidence that were vital to his thesis, and we will consider them in turn.

The evidence from breeding studies

It was well-known that offspring tend to resemble their parents more closely than other members of the species. The phenomenon was not understood, but it was established as an observational fact. At this time, the laws of heredity were still quite obscure, and people tended to believe that offspring were usually the result of a blending or mixing of the characteristics of the two parents.

But even if the mechanism was not fully, or even partly, understood, the fact of parent-offspring resemblance was used, and had been used for generations, by agriculturalists and plant or animal breeders. Consciously, or unconsciously, they had been choosing their best individuals and breeding from them, while discarding the less desirable forms. This had the long-term result of modifying and improving the strain, gradually changing its appearance towards the desired end. Darwin himself had performed selective breeding experiments with pigeons, and had successfully produced modification in the appearance of some of his stocks.

The problems of extinction

Darwin believed the observation that species often became extinct to be particularly relevant. The reasons why extinction takes place need not always be the same. If the environment should change and become unsuitable, a species might fail to survive; the implication being that it was unable to adapt to the changed conditions. Thus, Newell[204] has suggested that many species of marine invertebrates became extinct during the Devonian and Permian eras due to the reduction of the area of the seas and the appearance of the major continental land masses.

Alternatively, one species could become extinct as the direct, or indirect, result of the superior performance of another. The red squirrel (*Sciurus vulgaris*) is principally an inhabitant of coniferous forest and consequently many of the woodland areas of Britain are suboptimal habitats. When the grey squirrel (*S. carolinensis*) was introduced from North America, it spread very rapidly, and there was a correlated decline in red squirrel numbers, until the latter is now very much restricted in its range. This is good circumstantial evidence that the decline of the red squirrel was caused directly or indirectly by the presence of its congener.

Intraspecific competition

It is apparent from reading Darwin's books that he was greatly influenced by the writings of Malthus, and particularly by an essay which discussed the subject of population growth. Malthus himself was primarily concerned with the social problems of man in an urban environment following the industrial revolution. He appreciated that, without adequate controls, a population would increase rapidly and geometrically. In theory, if an average pair of individuals leave four offspring, then the population will double in size every generation. It will increase over a million fold in less than 25 generations. Such geometrical growth does not usually occur, although the increase in natural populations is sometimes remarkable.

For example, in 1938, two male and six female ring-necked pheasants (*Phasianus colchicus*) were released on Protection Island in the Puget Sound, northwestern U.S.A. Six years later, the population size was estimated at 1898 individuals by Einarsen.[85–6] This means that the population had increased over 200-fold, and the island is too isolated for immigration to be involved. The increase is due to the reproductive capacity of the birds themselves, and Lack[166] reports that the prime factor is a high survival of young birds. Presumably the density was so low that mortality due to intraspecific effects was quite unimportant, and the absence of competition allowed almost maximum survival and consequently a very rapid population growth. Unfortunately, the island was occupied by the military in 1942, and the population of pheasants crashed (into the pot!). The subsequent history is not available. In other species which have undergone rapid growth, the population size stabilized out at an upper limit, and remained reasonably constant at this level. Davidson[66] reports on the number of sheep in Tasmania between 1830 and 1925. The population rose from 200 000 in 1820 to 2 million in 1850. It remained at this level for the next 70 years. It could be argued that this is not a particularly impressive example since it is subject to human control, but it could also be argued that man is exerting a level of predation which is maintaining the sheep population at or about the 2 million mark.

EVOLUTION BY NATURAL SELECTION

Populations introduced into a reasonably stable environment do not usually increase geometrically for more than a few generations. Eventually they level out at some point, perhaps fluctuating somewhat from generation to generation. Unless the reproductive strategy of the species changes, such a levelling out of the population curve implies that mortality must increase. A higher proportion of the population fails to survive to maturity in a stable situation, than in one where numbers are increasing.

Darwin believed that these ideas of Malthus were of fundamental importance in evolution. He suggested that mortality (which regulated population size) took place in such a way that the less well adapted would be the victims. This does not necessarily imply that the physically weak die, rather that those individuals which utilize the general biological environment most efficiently will be the survivors. Darwin accounted for evolutionary change through this system. Just as a breeder chooses those individuals closest to his desired optimum, and discards the rest, so the natural environment improves the performance of a species by eliminating the less effective. Individuals possessing particular adaptations will survive better, and by virtue of the heritable nature of these adaptations, they will transmit them to their offspring. Gradually, the adaptations will spread and improve, so that the species will become better suited to the environment which it inhabits.

This is the fundamental principle of evolution by natural selection, and deserves repetition. A population produces an excess of offspring and mortality occurs to reduce the numbers to a level which the environment can support. There is inherited variation for all those characters which affect the probability of survival in each individual. Those individuals that are least well adapted to the environment perish, and the survivors pass on to the next generation the potentiality of producing the very characters which allowed their survival. Thus, the nature of the population changes as a result of this natural selection: the population evolves towards a new level of adaptedness to the environment.

Let us consider this theory in relation to the finches of the Galapagos Islands. A flock of ancestral birds arrived on the islands from South America in the distant past, and found them ornithologically barren. The flock presumably settled on one island, and then indulged in the exponential growth demonstrated by Einarsen's pheasants. When that island became fully populated, and mortality increased, adjacent islands were perhaps colonized one after another until the entire Archipelago was occupied. Gradually, the island populations were evolving under natural selection, accumulating differences which improved their success on their natal island, until they reached the level of full species. Secondary recolonization of the islands from which they came could then give rise to the complex finch fauna, which was apparent to Darwin and still exists today.

There are gaps in this story, however, which classical Darwinism cannot fill. For example, how does speciation actually occur? Is it anatomical, behavioural, or ecological? What is it that delimits one species from another? Darwin[64] himself dealt with the problem by suggesting that species was a term: 'arbitrarily given for the sake of convenience to a set of individuals closely resembling one another' and that the term 'does not essentially differ from the term variety which is given to less distinct and more fluctuating forms.'

However, Darwin is not really correct here. A species is perhaps the only taxon in systematic classification which is not arbitrary. It is allowable of a much more precise definition, either genetically as 'the largest and most inclusive reproductive community of sexual and cross-fertilizing individuals which share in a common gene pool',[76] or somewhat more biologically as 'groups of actually or potentially inbreeding natural populations which are reproductively isolated from other such groups.'[183] The important facts are that reproductive isolation exists between species, and that the act of speciation consists in its barest essentials of the establishing of such isolation.

Despite the title of his major work being 'On the Origin of Species', Charles Darwin did not actually discuss this to any great extent; he was merely concerned with establishing the fact of evolution and his thesis that it took place through the action of natural selection. Because of his loose definition of the status of a species, he did not have a problem in accounting for its origin. It is merely a grade of difference between groups of organisms that arises by virtue of natural selection acting upon the variation present in an earlier series of populations. However, we now regard a species as something more precise, and will return to the phenomenon of speciation later in this book.

THE CONTRIBUTION OF ALFRED RUSSELL WALLACE

The name of Charles Darwin must always be associated with that of Alfred Russell Wallace in discussion of the theory of evolution by natural selection, for both independently derived the theory during the mid-nineteenth century. Wallace was a very different man to Darwin, being less well educated, and a professional collector rather than a gentleman scientist. Sheppard[239] points out that the two men shared the attributes of being field naturalists of great experience and ability, with a lack of formal instruction in biology. This lack of training enabled them to view the subject with an eye unjaundiced by the dogma of the age.

The remarkable adaptations of organisms to their environment was as apparent to Wallace[264] as it was to Darwin, and the former similarly appreciated that a species produced far more offspring in a generation than the environment could support. He believed that there would be tremendous competition for food in most natural populations, and that its availability would be the principal factor controlling their size. While this is undoubtedly true in some situations, other factors can be equally important, and can be selective agents under alternative circumstances.

Evidence in support of the relevance of food supply to population dynamics has been provided by Lack,[166] with a particularly impressive result concerning the survival of nesting swifts in an Oxfordshire colony. Lack demonstrated that mortality varies considerably from year to year, and Table 1.2 shows that it is greatest in seasons when the weather is poor. He suggested that this is related to the availability of aerial insects upon which adult swifts feed their young. In summers when the weather is fine, there are plenty of insects and so the nestlings get sufficient food. In poor seasons, insect food is less readily available, and mortality is higher.

The climatic situation shows an interaction with clutch size, however, which is both interesting and instructive. Swifts usually lay either two or three eggs in a clutch. As can be seen from Table 1.2, when the breeding season coincides with fine weather and a plentiful food supply, an average of 2.3 young survive from nests containing

Table 1.2 The survival of nestling Swifts (*Apus apus*) in different years, related to the amount of sunshine, and hence to the availability of aerial insects upon which adults feed their young. From Lack.[166]

	Brood size	Number of young hatching	Proportion surviving to flying stage	Number of young raised per brood
1946 to 1948 sunshine well below average	2	48	50%	1·0
	3	36	31%	0·9
1949 to 1952 sunshine average or above average	2	156	95%	1·9
	3	60	75%	2·3

three eggs, but only 1.9 from nests with two. In poor seasons, the picture is slightly different: only 0.9 young survive from three-egg clutches, while 1.0 survive from those with two eggs. Here is an example of selection pressure imposed by food supply. In a good season, parents have no difficulty in rearing three youngsters, and leave more offspring than their two-egg neighbours. When the climatic conditions are more severe, a pair of swifts can collect enough food for two nestlings, but insufficient for three, and the whole brood starves. Consequently, more young survive from clutches of two in these years, and selection operates against parents who produce large clutches.

Food supply is not always as important as this. The possession of a territory is vital for many species of temperate song birds to be able to breed successfully, but there is no great evidence that territories are smaller in seasons when food is particularly abundant. Consequently, the densities of populations are not directly limited by food supply, although if food is locally superabundant, the survival of the nestlings may be enhanced. Wallace oversimplified the situation when he suggested that competition for food was paramount, but 100 years later, the relative importance of territory, predation, parasitism and available food supply is still a matter for considerable dispute (for a recent review, see Ricklefs).[227]

LAMARCK AND ACQUIRED CHARACTERS

The *sine qua non* of Darwin and Wallace's ideas on evolution by natural selection is the presence of inherited variation within a population. Unless a population is variable, there is nothing to select.

Unless the variation is inherited, selection will have no lasting effect upon a population. Just as an agriculturalist cannot improve his stock if all the individuals are identical, so evolution cannot take place without heritable differences between the members of a species or population.

Darwin was unable to account for the presence of this heritable variation in experimental and natural populations. He appreciated that it was there, but could neither explain why it arose, nor how it was maintained. He considered the ideas of the French biologist Lamarck, who had proposed an explanation which accounted for the appearance and spread of characters which adapted a species to a particular role. Darwin was clearly unhappy about Lamarck's views, but the lack of hard facts concerning inheritance led him to consider them, as we must also.

Lamarck's theories concerning evolutionary processes are not generally accepted today for a variety of reasons, but he had one vigorous apologist in H. G. Cannon[34] who summarized his views on evolution as follows: 'when an animal is urged to any particular actions, the organs which carry out this are immediately stimulated. The result is that the ... repetitions of these acts strengthen, extend, develop and even create the organs that are necessary.' Organs or adaptations which are needed by a species can thus be created, and by their inheritance, pass into the constitution of the species. This is the crux of Lamarckism. The development that takes place by need and use becomes incorporated into the heritable nature of the species. The distinction between Darwinism and Lamarckism is fine, yet fundamental. The former requires that the variation is already present, and describes its effect upon the evolution of a population. It does not account for the origin or maintenance of the variation, and here lay the attraction of Lamarckism, which offered a reason for the appearance of the very organs which were needed.

Experiments have been performed from time to time in an attempt to demonstrate the supposed importance of 'need' in the establishment of inherited characters. Almost without exception, the experiments have failed, or proved unrepeatable, or the design has been shown not to exclude the possibility of natural selection acting upon pre-existing variation. Perhaps the most widely known of these experiments were undertaken during the years around the First World War by the Austrian biologist Kammerer.[140, 141]

One series of experiments concerned the midwife toad (*Alytes obstetricans*). This is a curious animal, for toads typically return to the water to breed and possess aquatic tadpole larvae. The midwife toad is exceptional in being permanently terrestrial, and the male is adapted for carrying the eggs around its back legs. Now, a toad in water is somewhat slippery, and to facilitate copulation, the males of most species possess spiny processes and calluses on the hands to assist in gripping the female's body. The midwife toad mates on land, and so has a dry skin. Consequently, these nuptial pads are unnecessary, and are absent.

Kammerer maintained a group of midwife toads in water for several generations, and claimed that the hands of the males developed horny patches which were inherited in subsequent generations, his argument being that the male toads needed these pads to maintain their position during copulation. Were this true, it might be a significant finding, provided the experiment was designed carefully enough to exclude the possibility of selection acting upon slight variations in the nature of the hands which were already present in the population. However, subsequent efforts to repeat Kammerer's experiments have failed to produce the same results. In a recent book, Koestler[160] has reviewed the evidence that the pads might have been synthesized fraudulently, either by Kammerer or one of his associates. Be that as it may, the lack of reproducibility of the results is a serious problem. It is one of the bases of scientific research that an experiment can be repeated independently, and still give the same results. This methodology is clearly of supreme importance when the initial result is claimed to run counter to the mainstream of scientific thought.

Kammerer undertook a second series of experiments using sea squirts (ascidians). He cut off the siphons of some experimental animals, and allowed them to regenerate. The regenerated siphons were larger than the originals. This is not too surprising a result, for it is well-known that sea squirt siphons are longer in sheltered water than in situations where the animals are subjected to wave action. The laboratory conditions would undoubtedly be more sheltered than those from which the animals originated. Kammerer then bred from these regenerated sea squirts, and found that the longer siphons were inherited. He claimed that this was further evidence of the inheritance of an acquired character, but both generations had been influenced by the same novel environmental factors so their resemblance is not too surprising. As with the midwife toad, it is important that these experiments should be repeatable. Kammerer's results have not been obtained subsequently, and the plasticity of a character such as siphon length in sea squirts is so great that very careful and rigorous control is necessary to determine that the claimed result is real. In the absence of such rigour, we are perhaps justified in dismissing the experimental result, for the burden of proof lies with the appellant in a situation where a new theory is being proposed.

One could list other experiments which have been claimed to give support to the idea that a character that is acquired during development can become incorporated into the hereditary material of an individual. In all cases they seem to have proved to be unrepeatable or just plain wrong. The situation stands at present that there is no conclusive evidence that evolution can take place in this way.

MENDEL AND NEO-DARWINISM

The breakthrough needed by the Darwin-Wallace theory of evolution by natural selection came in 1900 when De Vries,[73] Correns[58] and van Tschermack[256] simultaneously rediscovered the laws of inheritance. All of them had been independently studying the inheritance of simple, discrete characters, and had quantified their results. They ignored complex characters such as size and shape, and concentrated their efforts upon easily distinguished and discontinuous factors. When they followed up the scientific literature, they found that essentially the same results and conclusions had been published in an obscure journal in 1866 by Gregor Mendel,[186] a virtually unknown monk in the monastery at Brunn.

These results were exactly the stimulus which was needed. The Darwinian theory was beginning to slip in esteem at the time because of the lack of backbone which an understanding of the nature and mechanism of inheritance would have given. In the years following 1900, resemblances between relatives became explicable and the similarities between parents and offspring were accountable through the possession of genes of similar origin. Furthermore, the variation between individuals within a population could now be explained. Discrete variation could be ascribed to the presence of segregating loci within a population, for homozygotes and heterozygotes might differ in their phenotypes. More complex characters could be explained in terms of several loci, each with segregating alleles. Differences between individuals in metric characters could be accounted for if they bore different series of alleles at these loci. Furthermore, the continuing segregation of the alleles would maintain the phenotypic variation from one generation to the next.

Equally importantly, the origin of the new variation could be defined. Mutation resulted in the change of one allele for another, and gave rise to a new phenotype. So the appearance of rare individuals could sometimes be explained in terms of the occurrence of a new mutation, or alternatively the entry of a new gene into the population through migrant individuals.

Following the rediscovery of Mendel's work, the theory of evolution by natural selection has been remodelled and revised, but without any altering of the basic underlying concepts. Organisms are still understood to overproduce their numbers, and random assortment of the chromosomes during meiosis coupled with recombination ensures that genetic variation is manifest in every population. If a particular characteristic is favoured in a given environment, the genes giving rise to it survive the selection process and are passed on to the next generation. In this way selectively advantageous alleles increase in frequency at the expense of less suitable ones, and the genetic constitution of a population or species changes and evolves.

Neo-Darwinians, as they came to be called, have given rise to more precise definitions of species, populations and variation, both individual and between populations. These topics come together within the domain of population genetics, which, with its environmental associate ecological genetics, now forms the core of contemporary studies into the theory and mechanism of evolutionary change at the level of the individual gene.

The following chapters will attempt a review of the situation in some aspects of population genetics that relate directly to the evolutionary process. We shall draw together studies that are now regarded as classical, and others that no doubt will become so. The goal will be to provide a simple yet comprehensive review of the evidence, both experimental and theoretical, relating to evolutionary change in populations and species through the substitution of one gene for another. We shall attempt to show how gene replacement can be influenced either by selection, more or less as envisaged by Darwin, or perhaps by other forces. We shall see how this can lead to the divergence of populations, and ultimately to the establishment of new species.

2

The Behaviour of Genes in Populations

INTRODUCTION

We have seen in the previous chapter that species usually evolve by the gradual accumulation of differences in morphology, ecology, physiology or behaviour until reproductive isolation has occurred. The crucial stage of this isolation must be when the genomes of two incipient species have become sufficiently different for chromosomal pairing and regular disjunction to be impossible at meiosis. This results in the hybrid being sterile, for it cannot produce gametes and there is consequently a reproductive barrier between the two populations.

This differentiation of the genome takes place by the reorganization of the genetic material, either suddenly, following major genetic dissumptions such as large inversions or polyploidy, or gradually by the incorporation of many individual small differences. It is the latter process with which we are now concerned. Over the course of evolutionary time new alleles are produced in certain individuals. Some of these alleles increase slowly in frequency and eventually become established in the population or the species. The ultimate source of much of this new material is mutation. It does not matter to the mathematical theory of evolutionary genetics how the mutation occurs; the fact that rare events produce new or modified bits of DNA is sufficient. This new material will usually arise as a mutant gene in a single individual and die out within a few generations. Occasionally, a mutation will spread until all members of the population carry it. The way in which this may happen is a matter of great importance to evolutionary genetics and forms the subject of the present chapter.

GENE-FREQUENCY

The study of the distribution and behaviour of alleles within a population is part of population genetics. This has largely been a theoretical subject, although in recent years increasing numbers of laboratory experiments have been undertaken, usually involving simplified situations where the maximum possible number of variables have been controlled. There is an ecological branch of the science which deals more particularly with the behaviour of alleles in nature. Because of the vagaries of the outside world, it is often difficult, if not impossible, to control its variables, and so ecological genetics is a more statistical and less precise science. Both of them, however, are concerned with the distribution of genes in populations, and their common basic parameter is 'gene-frequency'.

The simplest examples of gene-frequency stem from genetic polymorphism. This was defined by Ford[100] as the occurrence together in the same locality of two or more discontinuous forms of a species, in such proportions that the rarest of them cannot be maintained merely by recurrent mutation. This definition excludes geographic variation from population to population. For example, in the bluethroat (*Cyanosylvia svecica*), adult males possess a blue throat in the centre of which is a conspicuous spot. In Scandinavian populations, the spot is red and the bird is described as the subspecies (*C.s. svecica*) distinct from the white-spotted form (*C.s. cyanecula*) of eastern and central Europe. Rarely, if ever, do the two forms co-exist on the breeding grounds, and so they cannot be regarded as polymorphic forms.

Ford's definition also excludes situations such as the disease of man called Hunter's Syndrome or Gargoylism which has only been recorded in about 150 families in the whole world,[49] and is usually inherited as an autosomal recessive. Levels of incidence as low as this may be the result of recurrent mutations and the chance marriage of extremely rare heterozygotes.

A much more widespread human variation involves the MN blood groups. Three phenotypes can be identified by agglutination tests, and they are called M, MN and N. Genetic analysis of families indicates that two co-dominant alleles (M and N) are responsible, segregating at a single locus. Thus, the corresponding genotypes are MM, MN and NN. Table 2.1 lists the composition of samples of people taken randomly from six different races of man. In the sample of U.S. whites, there were 1787 MM individuals, 3039 MN's and 1303 NN's, giving a total of 6129 people tested. Such a level of incidence of the three MN genotypes is sufficiently high for this to be

Table 2.1 The frequencies of MN blood groups in different populations. Taken from Stern, C. (1960). *Principles of Human Genetics*, p. 158.

Population		MM	MN	NN	Total	Freq. (M)	Freq. (N)
U.S. Whites	Observed Number	1787	3039	1303	6129		
	Expected Proportion	0.2916	0·4968	0·2116		0·540	0·460
	Expected Number	1787·2	3044·9	1296·9			
U.S. Negroes	O.N.	79	138	61	278		
	E.P.	0·2835	0·4989	0·2186		0·532	0·468
	E.N.	78·8	138·7	60·8			
U.S. (Red) Indians	O.N.	123	72	10	205		
	E.P.	0·6015	0·3481	0·0504		0·776	0·224
	E.N.	123·3	71·4	10·3			
East Greenland Eskimos	O.N.	475	89	5	569		
	E.P.	0·8335	0·1589	0·0076		0·913	0·087
	E.N.	474·3	90·4	4·3			
Ainus (Japan)	O.N.	90	253	161	504		
	E.P.	0·1845	0·4901	0·3234		0·430	0·570
	E.N.	93·0	247·0	163·0			
Australian Aborigines	O.N.	22	216	492	730		
	E.P.	0·0317	0·2926	0·6757		0·178	0·882
	E.N.	23·1	213·6	493·3			

U.S. Whites: Freq. of M = (1787 × 2 + 3039)/(2 × 6129) = 0·540 = p
Freq. of N = (1303 × 2 + 3039)/(2 × 6129) = 0·460 = q

Expected proportion of MM = p^2 = $(0{\cdot}540)^2$ = 0·2916
,, ,, ,, MN = $2pq$ = 2(0·540)(0·460) = 0·4968
,, ,, ,, NN = q^2 = $(0{\cdot}460)^2$ = 0·2116

These are obtained from the Hardy-Weinberg Law.

Expected number of MM = p^2N = 0·2916 × 6129 = 1787·2
,, ,, ,, MN = $2pq$N = 0·4968 × 6129 = 3044·9
,, ,, ,, NN = q^2N = 0·2116 × 6129 = 1296·9

a true genetic polymorphism as defined by Ford.

We can use these data to illustrate the estimation of gene frequency for the M and N alleles. Every individual who is of the M phenotype must be homozygous MM in genotype, and consequently possesses two M genes. MN individuals, being heterozygous, possess one, and the remaining NN's have none. The number of M genes can therefore be calculated directly, simply by adding the number present in every individual, as follows:

$$\begin{aligned}\text{the number of M genes} &= 2 \times 1787 + 1 \times 3039 + 0 \times 1303 \\ &= 6613\end{aligned}$$

similarly,

$$\begin{aligned}\text{the number of N genes} &= 0 \times 1787 + 1 \times 3039 + 2 \times 1303 \\ &= 5645\end{aligned}$$

The total number of genes is 12258 which, of course, equals twice the sample size, because every person carries two alleles at the MN locus. We can now derive the frequency of the M and N alleles directly as the *proportion* of each in the sample.

The frequency of M $= 6613/12258 = 0.540$
and the frequency of N $= 5645/12258 = 0.460$.

There are only two alleles at this locus in the sample, and consequently the two gene-frequencies must add up to one.

It is important to appreciate that these calculations are based upon a sample of individuals, and consequently only give an *estimate* of the gene-frequency in the population at large. As with all estimates this carries an error, and, provided the sample is large and the population even larger, the error approximates to

$$\sqrt{\frac{p(1-p)}{n}}$$

where p is the frequency estimated from a sample size n. For the M allele in the sample described above the error is

$$\sqrt{\frac{0.540 \times 0.460}{6129}} = 0.0064$$

It is clear from this equation that an error will decrease as the sample upon which the frequency is based increases. Logically, this must be so. A larger sample will yield a more reliable estimate of gene-frequency, and so the error will decline.

We can now turn to the question of how alleles are distributed within a population. To simplify the situation, we will begin with a consideration of the behaviour of alleles in a non-evolving population, unaffected by the disturbing influence of mutation, migration or natural selection. We will then turn to the more critical situation where these forces are acting upon a population, and the individuals which compose it.

THE HARDY-WEINBERG LAW

The law which describes the relationships between gene-frequencies and genotype frequencies in a random mating population was discovered more or less simultaneously by a series of workers in the early years of the present century. It now usually bears the names of two of them, Hardy and Weinberg, whose papers on the subject were published in 1906. Hardy was a mathematician of some repute, and was rather embarrassed to find that a law of such mathematical simplicity was unknown to the geneticists of the day. It is now a fundamental law of population and hence of theoretical evolutionary genetics, and forms the cornerstone for the remainder of this chapter.

We will consider first the situation where two alleles A_1 and A_2 are segregating at an autosomal locus in a large, randomly mating population. There are consequently three genotypes A_1A_1, A_1A_2 and A_2A_2 which we will suppose to be identical in all biological parameters such as behaviour, fertility and longevity. Suppose also that the species is hermaphrodite, and that self-fertilization is as likely as fertilization by any other member of the population. Finally, for simplicity, let us suppose that the generations are discrete: that is, all the parents die before their offspring reach reproductive age, and so take no part in the production of any subsequent generation.

Let the number of individuals in the population at a particular time be as follows:

$$\begin{aligned} \text{Number of } A_1A_1 &= D \\ \text{Number of } A_1A_2 &= H \\ \text{Number of } A_2A_2 &= R \\ \text{Total number} &= N \end{aligned}$$

The proportion of the three genotypes will be:

$$\begin{aligned} \text{Proportion of } A_1A_1 &= D/N = d \\ \text{Proportion of } A_1A_2 &= H/N = h \\ \text{Proportion of } A_2A_2 &= R/N = r \end{aligned}$$

Let the frequency of the A_1 gene and the A_2 gene in the population be p and q respectively.

Then $\quad p = (2D+H)/2N = d+h/2$

and $\quad q = (2R+H)/2N = r+h/2$

Since mating is random with respect to genotype, the chances of a particular individual which is involved in the mating process being A_1A_1 is d. The probability that the mate is also A_1A_1 is d as well. By the multiplication law for combining probabilities, the proportion of matings which are A_1A_1 by A_1A_1 is given by the product of the individual probabilities—that is, by $d \times d = d^2$. We can estimate the frequency of all the mating pairs in exactly the same way, as shown in Table 2.1.

Since we assumed that fertilities were all the same and that every individual had an equal chance of survival, all of these pairs will produce an equivalent number of offspring. From Mendel's laws we can write down the proportion of offspring of each genotype which will be produced by every pair. For example, the matings between two A_1A_2 individuals will produce offspring of the three genotypes in the ratio of 1:2:1. Since the frequency of such mating is h^2, the proportion of offspring will be $h^2/4$ A_1A_1, $h^2/2$ A_1A_2 and $h^2/4$ A_2A_2.

Table 2.2 This shows the frequency of all possible mating pairs in a large random-mating population. The final three columns indicate the frequency of genotype produced by these matings. For further details, please refer to the text.

MALE		FEMALE		PAIR	OFFSPRING FREQUENCY		
Genotype	Frequency	Genotype	Frequency	Frequency	A_1A_1	A_1A_2	A_2A_2
A_1A_1	d	A_1A_1	d	d^2	d^2		
A_1A_1	d	A_1A_2	h	dh	$dh/2$	$dh/2$	
A_1A_1	d	A_2A_2	r	dr		dr	
A_1A_2	h	A_1A_1	d	dh	$dh/2$	$dh/2$	
A_1A_2	h	A_1A_2	h	h^2	$h^2/4$	$h^2/2$	$h^2/4$
A_1A_2	h	A_2A_2	r	hr		$hr/2$	$hr/2$
A_2A_2	r	A_1A_1	d	dr		dr	
A_2A_2	r	A_1A_2	h	hr		$hr/2$	$hr/2$
A_2A_2	r	A_2A_2	r	r^2			r^2

Table 2.2 shows these proportions for all matings, and since survivals are all equal we can add up the proportions in each offspring genotype to find the totals.

$$\begin{aligned} \text{Total of } A_1A_1 \text{ offspring} &= d^2 + dh + h^2/4 \\ &= (d + h/2)^2 = p^2 \\ \text{Total of } A_1A_2 \text{ offspring} &= 2dr + hr + dh + h^2/2 \\ &= 2\left(d + \frac{h}{2}\right)\left(r + \frac{h}{2}\right) = 2pq \end{aligned}$$

$$\text{Total of } A_2A_2 \text{ offspring} = r^2 + hr + h^2/4$$
$$= \left(r + \frac{h}{2}\right)^2 = q^2$$

From this, we can easily derive the frequency of the A_1 gene among the offspring. The A_1A_1 individuals have two A_1 genes and the A_1A_2's have one. Consequently, the frequency is:

$$\frac{2p^2 + 2pq}{2p^2 + 4pq + 2q^2} = \frac{2p(p+q)}{2(p+q)^2}$$

Since $p+q = 1$, this reduces to p, and the frequency of the A_2 gene can be similarly calculated as q.

We have just derived two results which are of fundamental importance to theoretical evolutionary genetics. Firstly, the genotypes in a population occur in proportions which depend solely upon the gene-frequencies in the parental population. The relative numbers of parental genotypes are irrelevant. Provided that the conditions that we stated are adhered to, the relationship between the offspring genotypes will be $p^2:2pq:q^2$. This relationship can be extended to three or more alleles, giving the genotype frequencies as $p^2:2pq:2pr:q^2:2qr:r^2$ (where r is the frequency of the third allele), and so on.

Secondly, the gene-frequency in the offspring population is identical with that of its parents. This is a less surprising result. We built into the model the condition that there should be no differences between the genotypes in any biological parameters such as viability. Consequently, there is no reason why one gene should be over-represented in a subsequent generation. In the absence of any disturbing force, we would expect a population to remain in this state of equilibrium indefinitely.

The model that we have just discussed is the simplest possible, and assumed a randomly mating population of hermaphrodites. Altering the situation to allow for separate sexes, with or without different initial gene-frequencies, or non-overlapping generations, does not affect the outcome. The algebra becomes more complex, but the end result is the same. If we assume a sex-linked locus, the situation alters slightly, for several generations are required to reach equilibrium. It is not proposed to repeat the computations here. Interested readers are invited to perform the calculations for themselves, or seek it in a more specialized text such as Li.[177]

We are now in a position to make an important deduction from these two results. Since the genotypes in a population are present in frequencies that depend upon the parental gene-frequencies, and since these gene-frequencies do not change from one generation to the next, there should be the same simple relationship between gene and genotype within a single generation. If we look, and find that this is so, then we have some evidence that real populations can behave in the way suggested by our simple model.

The most impressive data for looking at this have been provided by Stern[251] concerning the MN blood groups in man. There are two alleles, M and N, which are co-dominant, so the heterozygous condition can be recognized during serological testing. Table 2.1 lists some of Stern's data. The number of individuals of each blood group are given from samples taken from six different races of man. As we have seen, it is a simple matter to estimate the frequency of the M and N alleles. These can be substituted in $p^2:2pq:q^2$ to give estimates of the genotype frequencies. Multiplying each of these frequencies by the number of people in the sample gives expected numbers of each genotyped. It does not require sophisticated statistics to show that they agree remarkably well in all the samples listed, despite the large differences in gene-frequency between the races.

Intuitively, it seems likely that human races give a fairly close approximation to a large population. Only the eccentric choose their spouse with regard to the MN blood groups, so mating is probably random for this character, and there is no evidence of important differences in fitness between these three genotypes. Hence, the conditions that we imposed in the derivation of the model are fulfilled, and the distribution of genes and genotypes appear to be in agreement with our predictions.

One sometimes comes across populations that appear to deviate from the Hardy-Weinberg equilibrium. Such populations are of great interest, for they may result from non-random mating or differences in survival of the genotypes, which could be of considerable evolutionary importance. They must be treated with caution, however, for they could also be due to poor sampling technique, or differences at loci other than those under observation.

RANDOM CHANGES IN GENE-FREQUENCY

Random genetic drift

The simple model in the previous section suggested that gene-frequencies would not change from one generation to the next. However, this is not true unless the population is very large. The members of a population who actually participate in the reproductive process are effectively a sample of that population. In the derivation of the Hardy-Weinberg law, we assumed that it was a random sample and that it was identical in its composition with the population from which it came. However, when a small random sample is taken from a large group of individuals, one does not expect the sample to have exactly the same composition as the group. We can see this in a simple example.

Suppose 200 balls are taken at random from a bag containing 250 each of red and white balls. The most likely result is that there will be 100 red and 100 white balls in the sample, but the probability of obtaining this particular result is only 7·3%. Although it is the most likely single result, there are so many other possibilities that a deviation from equality is much more probable.

The larger the sample, the closer it will be to the true population: the extreme result occurring when the sample includes all the balls in the bag. Conversely, the smaller the sample, the more likely it is to deviate by a given amount from the proportions in the bag. A sample of 20 balls containing 60% reds has a probability of less than 1%.

A similar situation occurs in natural populations. If the breeding adults in a population are only a few score in number, a slight excess of one genotype will affect the gene-frequency much more than in a group of several thousand. A smaller population of parents will be likely to deviate more severely from the frequency in the previous generation, but whatever the number of breeding individuals the gene-frequency among them will virtually never be identical with that in the population from which they come. The gene-frequency in the offspring will consequently differ from their parental population. Furthermore, after a change in the gene-frequency has taken place in this way, no matter how small it may be, there need be no restoring force to change the frequency back. The situation is analogous to neutral equilibrium in physics, as typified by a billiard ball on a perfectly flat table. A gentle push will move the ball to a new position where it will remain until another disturbance. If these pushes are random in direction, the ball will move to and fro with each disturbance until it finally falls into a pocket. Similarly in a population, slight accidents of sampling will result in a small change of gene-frequency each generation, until eventually one allele is lost. The other allele is then the sole representative in the population, and is said to be 'fixed'.

This process is called 'random genetic drift' because the gene-frequencies drift up and down following chance differences in survival or breeding of the individuals involved. When the number of breeding individuals is small, the fluctuations are relatively large, and the population reaches fixation more rapidly. It then remains in this fixed condition until mutation or immigration brings in a new allele, and the process can start again.

Inbreeding

We have seen the effects of taking a sample of individuals from a large population to form the parents of the next generation. We must now turn to the effects of the smallness of the sample itself. Not only is there a chance of deviating somewhat from the true composition of the population, but, when the number of breeding individuals is small, there is an increased likelihood of matings taking place between relatives. If two siblings mate there is a possibility of each passing on a copy of the identical parental gene (see Fig. 2.1). Such alleles are said to be 'identical by descent', and identical homozygotes may be

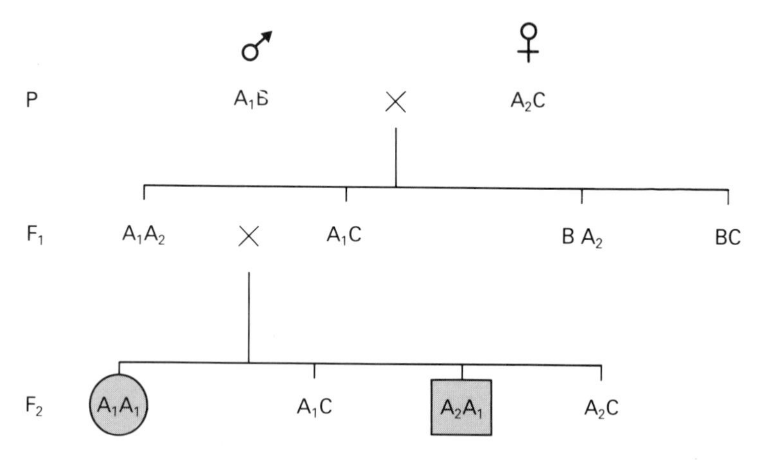

Fig. 2.1 In this pedigree, there are three distinguishable alleles, A, B and C. Both the individuals in the P generation are heterozygous for the A allele. These two alleles are designated A_1 and A_2 for clarity. Six sib-matings are possible between the F_1 individuals of which only one is illustrated. It can produce 2 AA homozygotes. One at these A_1A_2 (squared) contains an A allele descended from each of the P_1 individuals. The other A_1A_1 (circled) carries A alleles descended from the *same* A allele in the P_1 male. This individual is said to be an 'identical homozygote' as it carries two alleles that are 'identical by descent'.

produced whenever relatives mate. Distant relatives are, however, less likely to do so than close ones. Thus, the continued mating of relatives in a small population results in an accumulation of identical homozygotes until eventually all members carry the same gene at a locus. Genetic variability is lost at a rate which depends upon the population size, and separate populations will diverge as a result of the fixation of different alleles.

Whatever the size of a population, not all members contribute to the next generation. Some individuals die before reaching the reproductive state. Some are not healthy or well enough nourished to reproduce despite their survival. Some may produce offspring which themselves fail to survive. The number that actually contributes to the next generation may be very much less than the population measured by conventional biological techniques. Furthermore, if there is an excess of one sex, every one of the rarer sex will contribute relatively more to the next generation.

If all of these difficulties were incorporated into a genetical model, it would rapidly become extremely complex. To simplify the situation, the concept of an 'effective population size' has been introduced into the theory. An effective population size assumes (i) that there are equal numbers of males and females in the population, (ii) that they mate at random, and (iii) that every individual makes a similar contribution to the offspring generation. The effective population size is usually less than the true population size, N. It is always closer to the number of the rarer sex.[177]

If we consider a population whose effective population size is N_e, there are $2N_e$ genes present at a given locus. Assume that mating is random and that self-fertilization is equiprobable with fertilization by any other individual. The probability that two gametes from the same individual fuse to form a zygote is $1/N_e$; the probability that they carry identical genes is $1/2N_e$.

We can use this to estimate the amount of inbreeding. The expression $1/2N_e$, which defines the probability that any individual is an identical homozygote, can be used as an inbreeding coefficient. If this is large, then the population is inbred. There is, however, an additional component to the coefficient because some of the $(1 - 1/2N_e)$ individuals which are *not* newly formed identical honozygotes, may be identical as a result of inbreeding in an earlier generation. We have defined the coefficient as the probability that an individual is an identical homozygote, so, in generation t,

$$F_t = \frac{1}{2N_e} + \left(1 - \frac{1}{2N_e}\right) F_{t-1}$$

This is a recurrence equation, and given a starting value of F, the value of the inbreeding coefficient can be computed for any number of subsequent generations. Conventionally, the inbreeding coefficient of the original population is set at zero, and Fig. 2.2 shows how much more rapidly a population becomes genetically identical at a locus when its size is small.

The immigration of individuals from outside into the ideal popu-

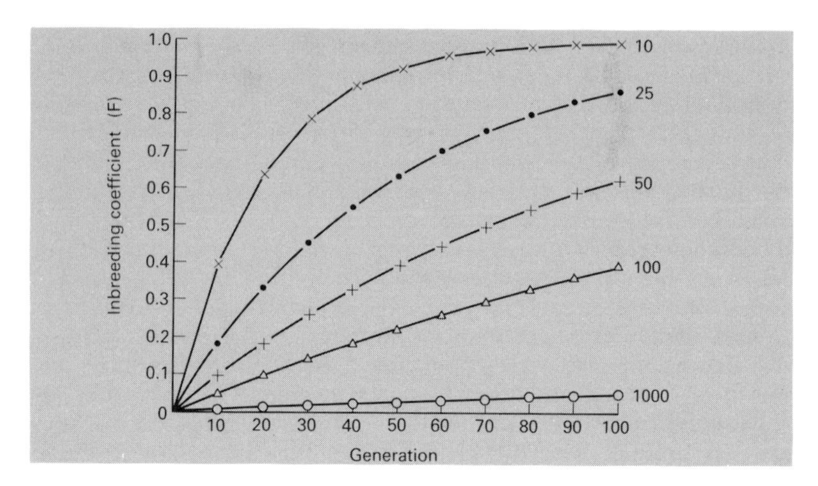

Fig. 2.2 The relationship between the inbreeding coefficient and generation number for a pair of alleles at a neutral locus in populations of various sizes. Notice that the coefficient increases much more rapidly when the population is small.

lation will result in the introduction of genes which are unlikely to be identical. If the immigration rate is M per generation, only $(1-M)$ of the population can carry native genes. Hence only $(1-M)^2$ of the zygotes can derive from the fusion of locally produced gametes, and the value of F_t must be reduced accordingly to:

$$F_t = \frac{1}{2N_e} + \left(1 - \frac{1}{2N_e}\right) F_{t-1}(1-M)^2$$

There will presumably be an equilibrium situation where the rate at which identical homozygotes are produced by inbreeding is balanced by immigration introducing non-identical alleles. At this point there will be no change in the amount of inbreeding, and so the coefficient will be constant at $\hat{F}$. Thus:

$$\hat{F} = \frac{1}{2N_e} + \left(1 - \frac{1}{2N_e}\right) \hat{F} \left(1-M\right)^2$$

which reduces to:

$$\hat{F} = \frac{1}{1 + 4N_eM - 2M - 2N_eM^2 + M^2}$$

To be able to take this equation further it is necessary to introduce one of the approximations so beloved by population geneticists. The immigration rate M is probably quite small, in which case M^2 will be negligible when compared to $1-2M$. So terms in M^2 can fairly safely be omitted to give:

$$\hat{F} = \frac{1}{1+4N_eM-2M}$$

which in turn can be approximated to:

$$\hat{F} = \frac{1}{1+4N_eM} \text{ since } 2M \text{ is small compared with } 1.$$

This equation is much simpler, and very interesting. N_eM is equal to the number of immigrant individuals in a generation, and the equilibrium coefficient depends upon this and nothing else. So whether the population is 100 or 100 000, if five strangers enter the population in each generation, $\hat{F} < 5\%$. And an inbreeding coefficient as small as this is sufficient to prevent populations from diverging. Thus, inbreeding and other random events will only be of importance in effecting a change in the genetic constitution of a population when migration between adjacent populations is very small, otherwise there will effectively be only one, very large, population.

The importance of random processes such as inbreeding and genetic drift to evolution has been a matter for dispute for a long time. Before we can review the evidence, we must consider other forces which can change the frequencies of alleles in natural populations. We will therefore defer a discussion of their evolutionary significance until later in the chapter.

DIRECTED CHANGES IN GENE-FREQUENCY

Mutation

Mutation is the ultimate source of new genes, and in our derivation of the Hardy-Weinberg law we ignored its effects. However, it is a process that is continuously acting and we must examine its effects upon the distribution of genes in populations. The mechanism of mutation need not concern us here. It is covered well by Lewis and John[174] and as far as the theoretical laws of population and evolutionary genetics are concerned, it does not matter how a new gene is produced. At present, we are solely concerned with the effects of its arrival upon a population, and shall regard a mutational event as one that changes one allele into another.

Suppose that in an ideal population, there is an allele A_1 which has a frequency p. All other alleles are called A_2, and have a frequency q, so that $p+q=1$. Let us suppose that mutation only occurs in one direction, changing A_1 to one of the alleles included in A_2. There will clearly be a progressive decline in the frequency of A_1. We can be more precise than this, and, if the mutation rate is u per gamete per generation, the frequency of A_1 in the next generation will be $p(1-u)$. In the subsequent generation it will be $p(1-u)^2$, and in the nth generation it will be $p(1-u)^n$. We can now estimate the length of time for the gene-frequency to change by a particular amount. Let t be the number of generations required for the frequency of A_1 to decline to $p/2$, then:

$$p/2 = p(1-10^{-6})^t$$

from which $t = 693147$. And this is true whatever the value of p: it takes nearly 700 000 generations to halve a gene-frequency under a mutation pressure of 10^{-6}.

Studies of natural populations show that gene-frequencies change much more rapidly than this. The peppered moth (*Biston betularia*) occurs in two varieties, which are controlled by a pair of allelic genes. The dominant (*carbonaria*) gene confers a blackish pigment upon the otherwise black and grey speckled adult. Until 1897, the black form had not been recorded in the London area. By 1905, it had increased to 37% and by 1952–6 it had increased further to 90% of the population (data from Kettlewell[148]). This represents a decline in frequency of the *typica* gene from over 0.99 to less than 0.35 in about 60 generations. Substituting these in the above mutation rate equation gives:

$$0.35 = 0.99\,(1-u)^{60}$$

This can be solved to give $u = 0.0172$ which is far higher than any mutation rate that has been measured to date. In *Panaxia dominula*, another species of moth, the frequency of the *medionigra* gene (which also confers a blackish pigment to the adult) declined from 6.5% in 1945 to 1.1% in 1955 (data from Ford[102]). This would correspond to a mutation rate of 0.163, which again is unbelievably high. The change in gene-frequency took place at a time when the population was estimated every year, and was never less than 1000 adults. If mutation can be ruled out, and the population is too large for random processes to be important, we are left with natural selection as the only possible answer. However, before we move on to consider this, there is a further example of mutation which we must consider.

So far, we have only discussed the effects of mutation in a single direction, but it is known from studies on micro-organisms that mutation is a two-way process. An allele A_1 can mutate to A_2, and A_2 can change to A_1. The frequencies of these two processes need not be the same however, but, when an allele has increased to an appreciable frequency, its back-mutation to the ancestral allele must be considered.

Reverting to the earlier example, suppose that the mutation rate back from A_2 to A_1 is v. The frequency of A_1 will change in a single generation from p to $p-up+v(1-p)$. The change in frequency is given by:

$$\Delta p = up - v(1-p)$$

When the population reaches equilibrium, the changes in each direction will be balanced and $\Delta p = 0$. Hence:

$$up = v(1-p)$$

which gives:

$$\hat{p} = v/(u+v)$$

The equilibrium frequencies of two alleles that are segregating in a large population and that are influenced only by forward and back mutation thus depends upon the magnitudes of these mutation rates. It seems unlikely that mutation rates for the same alleles will differ between populations, and so one would expect that all large populations which are not greatly influenced by random effects should have the same equilibrium gene-frequencies at a particular locus. One only has to look at the blood group data in Table 2.1 to see that this is not the case, and many other similar results could be quoted. So, it seems that mutation alone is not the prime force in maintaining gene-frequencies in natural populations. We have seen that it is not the sole cause of changes in gene-frequency in some populations. So, while there is no doubt that mutation is the *sine qua non* of evolution, it may not be important in the statics and dynamics of population genetics. The patterns of gene-frequencies observed temporally and spatially in nature are too changing for mutation alone to be important.

Selection

Selection by man is a potent force in moulding the phenotypes of some organisms. All the breeds of dogs have been produced by man's attempts to improve their appearance or performance for a particular task—whether it be dragging a sledge across the arctic ice, sitting on a silken cushion, or driving foxes to a bloody death in a ditch. The effects of selection upon quantitative characters such as the oil content of maize will be deferred until later. Here we will discuss the power of selection to alter the frequency of an individual gene in a population.

An example of changing gene-frequencies has been reported by Reed and Reed.[223] They established populations of *Drosophila melanogaster*, with known frequencies of mutant and wild type alleles, in bottles. Food was provided regularly, and samples of flies were removed periodically to assess the frequencies of the various genotypes. Figure 2.3 shows the results of one of their experiments using the sex-linked recessive mutant gene producing white eyes. Male *Drosophila* have only one X chromosome. Consequently, whatever gene is present at any given locus on this chromosome will manifest itself in the males' phenotype whether it be dominant or recessive.

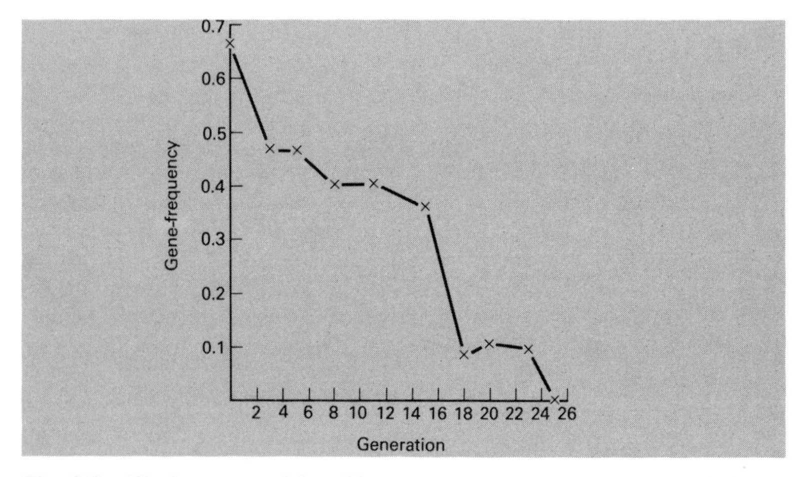

Fig. 2.3 The frequency of the white-eye gene in an experimental population of *Drosophila melanogaster* over 25 generations. Note the steady and progressive reduction to extinction at generation 25. Data from Reed and Reed.[223]

Thus, the proportion of males that had white eyes in Reed and Reed's experiments could be used as an estimate of gene-frequency in that sex. The frequency in the females, however, could only be estimated as the square root of the frequency of white-eyed flies among them. The gene-frequency in the whole population was estimated by weighting these two frequencies by the relative number of males and females in the population. Figure 2.3 shows that there was a steady decline in the frequency of the white allele.

Clearly, some differential effect is operating to reduce the frequency of the white-eyed flies. Natural selection is not an all-or-nothing effect, and although it is easiest to think in terms of individuals dying before they reach reproductive age, this is not the only way it can act. A reduction in fertility can be just as powerful a selective agent as biological death. In the case of the *D. melanogaster* experiment outlined above, Reed and Reed[223] showed that, in an illuminated bottle, white-eyed males are only 75% as likely to mate successfully, compared with their normal kin. Vision is an important component of the courtship behaviour of some *Drosophila* species. It seems that white-eyed males cannot see very well, and hence they mate less effectively, and suffer a consequent reduction of fitness. Reed and Reed showed that much of the decline in gene-frequency could be explained by this phenomenon.

It is possible to derive theoretical models to show how selection acts upon the genotypes of a locus. Their actual structure may be very complex, and take into account such factors as mating behaviour, differential fertility of the sexes, etc. Initially, however, we shall consider the simplest situation, where the genotypes exhibit differential mortality at some stage of the life cycle between zygosis and gametogenesis. We shall use the term 'fitness' to denote the probability of survival of a genotype relative to an arbitrary standard genotype, whose fitness is expressed as unity.

Consider two alleles A_1 and A_2 segregating in a large, randomly-mating population. We will suppose that the frequencies of the two alleles are p and q respectively, and set the fitness of the A_2A_2 genotype at $(1-s)$ relative to the other two genotypes, which are given a fitness of 1. Thus A_2 is treated as a recessive gene and the A_2A_2 genotype is at a selective disadvantage (represented by s, which can take any value from 0 to 1). If $s = 0$, the three genotypes have the same fitness; if $s = 1$, all the A_2A_2 individuals die before they reach reproductive age.

The relative proportions of the three genotypes after selection can be determined by multiplying the genotype frequency by its fitness as shown in Table 2.3. The frequency of the A_2 allele in the subsequent

Table 2.3 A simple model to demonstrate the effects of selection upon genes segregating at an autosomal locus.

Genotype	A_1A_1	A_1A_2	A_2A_2
Frequencies of the zygotes	p^2	$2pq$	q^2
Relative fitnesses of the genotypes	1	1	$1-s$
Proportions surviving to reproduce	p^2	$2pq$	$q^2(1-s)$

Total surviving the selection process $= p^2+2pq+q^2(1-s)$
$= 1-sq^2$

The frequency of the A_2 allele after selection $= q'$

$$= \frac{2pq+2q^2(1-s)}{2(1-sq^2)} = \frac{pq+q^2-sq^2}{1-sq^2}$$

Because $p = 1-q$, this simplifies to $q' = \dfrac{q-sq^2}{1-sq^2}$

generation (q') can then be determined by counting A_2 genes and dividing by twice the total number of genes:

$$q' = q - sq^2/1 - sq^2$$

We can then calculate the amount by which the frequency of A_2 changes in a single generation:

$$\begin{aligned}\Delta q &= q' - q \\ &= -sq^2(1-q)/(1-sq^2)\end{aligned}$$

If $\Delta q = 0$, there is no change in gene-frequency. Such situations are termed equilibria, and if an equilibrium arises because the locus is not segregating, it is called a 'trivial equilibrium', for there can be no change in gene-frequency under selection if only one allele is present in a population.

In the present model, Δq will be zero if $q = 0$, $1 - q = 0$, or $s = 0$. The former two of these are trivial points, and the third reflects the situation where all three fitnesses are equal. If s lies between 0 and 1, Δq will be negative. This is an intuitively obvious result: if a homozygote leaves fewer offspring, then the frequency of the gene which it carries will decline.

The equation derived above has particular relevance to the problem of controlling genetic disease in man. There are many debilitating conditions which are controlled by autosomal recessive genes. Alkaptoneuria, albinism and cystic fibrosis are perhaps the best known, although there are many others which cause equally distressing illnesses. The incidence of these diseases would clearly be reduced if affected persons were discouraged from producing children. If this could be done completely, the fitness of the affected homozygote would be reduced to zero, and the decrease in the gene-frequency would be the greatest possible. However, we have already noted that the change in gene-frequency is related to its frequency in the previous generation. Table 2.4 shows this relationship, and it is apparent that, when the incidence of the disease in the population at large is 1% or less, the change in frequency of the disease-causing gene declines very slowly. Indeed it is approximately equal to the incidence rate, and much less than the frequency of the gene. This is perhaps not very surprising, for, when a gene is rare, homozygotes will be extremely uncommon. It is precisely these individuals upon whom selection acts, and if they are rare, it can have little effect upon the overall frequency of the genes responsible. For example, the well-known recessive disease, cystic fibrosis, affects about one in 250 000 of Oriental and Negro people. Complete cessation of reproduction by individuals with the disease will only reduce the gene-frequency by 0.000 04 per generation, which equals one in 250 000. With rarer

Table 2.4 The relationship between the incidence of a disease in a population, the frequency of the recessive gene which causes it, and the change in frequency of that gene that would result if affected individuals were dissuaded from producing children. (All frequencies are expressed as percentages.) When such a condition is rare, it is noticeable that the change in the gene frequency is much lower than the frequency of the disease-causing gene. In fact, it is very close to the incidence of the disease within the population.

Incidence of disease	Frequency of gene	Decrease in gene frequency
90	94·9	46·2
80	89·4	42·2
70	83·7	38·1
60	77·5	33·8
50	70·7	29·3
40	63·2	24·5
30	54·8	19·4
20	44·7	13·8
10	31·6	7·6
5	22·4	4·1
1	10·0	0·9
0·1	3·2	0·09
0·01	1·0	0·01
0·001	0·3	0·001
0·0001	0·1	0·0001

diseases, the effects of selection are even less. This is the reason why so much research into human genetics is directed towards the identification of heterozygous, or 'carrier' individuals. Control of reproduction by these people would be far more effective in reducing the incidence of the disease.

Dominant genes, on the other hand, are much more susceptible to the effects of selection. It is easy to follow the method of Table 2.3, reversing the selection régime. If the fitness of the A_2A_2 genotype is set at unity, and that of the other two at $1-s$, the frequency of A_2 in the next generation is given by:

$$q' = (q - sq + sq^2)/(1 - s + sq^2)$$

Again, we can see the value of this equation by reference to problems in human genetics. Diseases such as one kind of muscular dystrophy and polydactyly are controlled by dominant genes. If we could persuade affected persons not to have children, their fitness would be reduced to zero. In this situation, $q' = 1$. A moment's reflection shows this to be self-evident. If no individuals carrying a dominant gene reproduce, then that gene will be eliminated from subsequent generations, apart from the occasional spontaneous mutant. Many of these dominant conditions show a reduction in fitness in normal populations, so natural selection is already acting to lower the gene-frequency. However, from the argument above, we can conclude that the control of reproduction might be of value in reducing the level of dominant genetic diseases even further.

THE FATE OF A MUTANT GENE

There can be little doubt that most newly arisen mutant genes are disadvantageous. The removal or insertion of a nucleotide into a gene will interfere with the triplet sequence from that point until the end of the gene. Mutations in which one base is substituted for another are less likely to disrupt the message (unless a nonsense codon is produced) but many of them will result in new triplets, coding for amino acids that are incompatible with the correct structure or function of the protein molecules specified by the gene.

It is not known what proportion of the remaining (non-disadvantageous) mutants are selectively neutral, and how many are advantageous. However, our considerations of the effects of sampling and of small population size made it clear that selectively neutral alleles *can* increase in a population until they are fixed. Indeed, as long ago as 1930 Fisher realized that this was the case and enquired what proportion of such mutants will spread to fixation. His calculations[96] suggested that the proportion is small: indeed, in an infinitely large population, no neutral allele will become fixed, and hence evolution by the substitution of neutral genes is impossible. But populations are not infinite, and after allowing for finite population sizes, Fisher suggested that there is a 79% chance that a new mutant allele which is selectively neutral will be lost in the first seven generations after its appearance.

More recently, Kimura[152] has examined this problem using more sophisticated mathematical techniques. He has proved that the probability of fixation of a neutral mutant is $1/2N_e$, where N_e is the effective size of the population in which the mutation occurred. This fraction is of course equal to the frequency of the mutant gene at the time of its appearance, and implies that only a tiny fraction of these mutations are successful. Kimura and Ohta[154] then went on to show that the average length of time taken by such a gene to spread to fixation is $4N_e$ generations. The genes which do not become fixed are ultimately lost, and the expression for the length of time this takes is more complex but simplifies to $2 \log_e(2N)$. The confidence limits of this expectation are large, for a neutral gene may drift out of the population at once, or increase to quite a high frequency before declining to extinction.

Let us now consider mutant genes which confer a selective advantage upon those individuals fortunate enough to carry them. We have already derived a model to describe the effects of selection upon a segregating population. There is a steady and progressive increase in the advantageous mutant allele until it is the sole representative in the population. Our model assumed a large population, but if the population is only moderately sized, random processes might be important. In fact, Fisher[96] showed that even if the population were infinite, with a selective advantage of 1%, the gene could still be lost by random processes in the first few generations. The probability that it would spread to fixation is only 2%.

Again using more sophisticated techniques, Kimura[149] showed that the probability of fixation is very complex, but that it simplifies to $2sN_e/N$. If N_e is equal to N, this reduces to twice the selective advantage, or $2s$, which is the result that Haldane[117-18] had also deduced by purely deterministic methods. This implies that if a mutation confers an advantage of 1% on the individuals which carry it, the probability of loss due to random effects is over 98%.

These results are of great significance to evolutionary genetics. We have already suggested that most mutants will be at a disadvantage, and be lost very rapidly either by drift or selection. Of the neutral alleles, most will be lost in the early generations, and only a very few will ultimately become fixed. Most of the advantageous alleles will also be lost in the early generations by drift, and only a few will increase sufficiently for selection to start showing its effect.

EXPERIMENTAL STUDIES ON GENE-FREQUENCIES IN SMALL POPULATIONS

From the theory outlined in the previous section, it is apparent that the genes that are found to be fixed in present-day populations may have spread through the species and replaced earlier alleles either selectively or neutrally. Distinguishing between these processes after the event is difficult. Indeed, it is only subsequent to 1965 that it has been possible to compare directly genes coding for the same protein in different species (see Chapter 3). As a result, it is only recently that it has been possible to show conclusively whether two species do in fact differ at all. Prior to 1965, efforts were directed towards the analysis of situations in which genes were segregating within a single population. These studies were directed towards determining whether the alleles concerned were spreading selectively or neutrally at the expense of others.

One line of research involved the use of experimental populations, and two of the most important contributions came from Kerr and Wright[144–5] who undertook a series of experiments with *Drosophila melanogaster*. In the first experiment, they established about 120 lines, using four males and four females as parents for each line. The lines were segregating for the sex-linked recessive gene 'forked', which modifies the shape of the bristles on the thorax. The initial frequency of the forked gene was 0.5 in every line. In each generation, the progeny were lightly etherized, and four males and four females chosen at random as the parents of the next generation. Kerr and Wright recorded when the lines became fixed for forked or wild type.

After 16 generations, they found that 29 lines had become fixed for the forked gene, and 41 for the wild type (see Fig. 2.4). There were 26 lines that were still segregating, and the remainder had become lost 'on account of mite infection and other accidents'. If the lines became

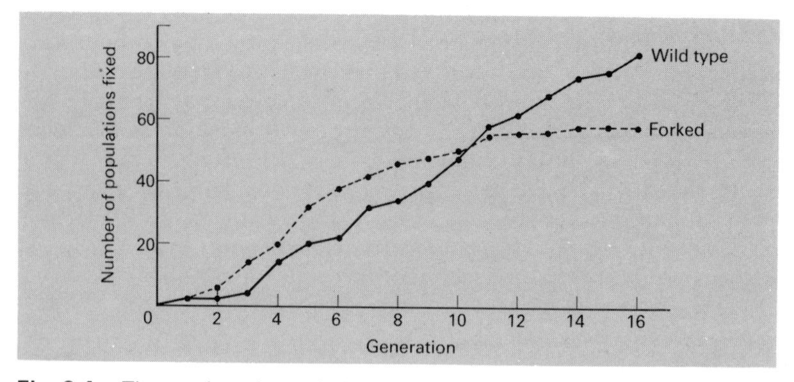

Fig. 2.4 The number of populations out of a total of 96 that had become fixed either for the wild type or the mutant gene in 16 generations of an experiment using very small populations. Data from Kerr and Wright.[144]

fixed randomly, we would expect an equal number to be fixed for each mutant. Kerr and Wright's results do not differ statistically from 35:35, and so there is no evidence that the wild type allele was fixed in more lines than the forked. Wright[277] had earlier calculated the rate of fixation for a neutral sex-linked gene. Kerr and Wright compared their results with the theoretical expectation, and found them to be in substantial agreement, which implies that genes really can spread through a small population under random genetic drift. In a second experiment,[145] using the Bar-eye mutant in the same species, the Bar flies were so disadvantageous that 88% of the lines had become fixed for wild type in 10 generations (Fig. 2.5). So, if it is strong enough, selection can overcome random drift, even in small populations.

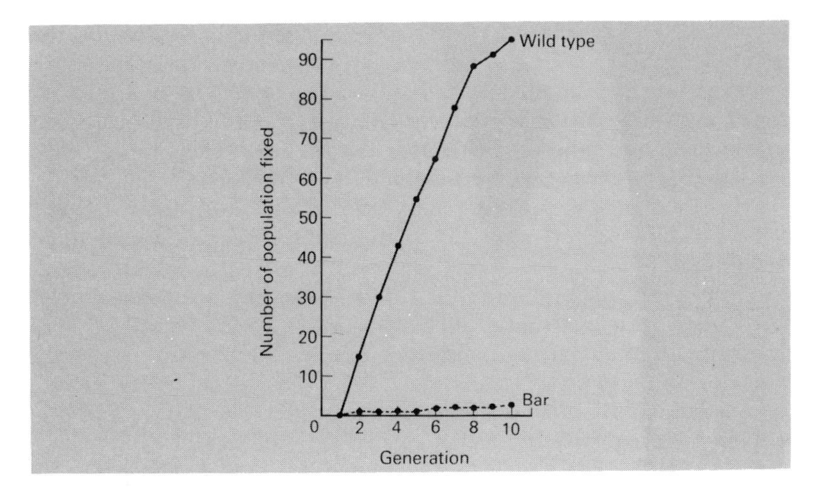

Fig. 2.5 The number of populations that had become fixed for the wild-type or for the mutant gene 'Bar' during 10 generations of an experiment using very small populations. Note that the vast majority of fixations relate to the wild type allele compared with the situation in Fig. 2.4. Data from Kerr and Wright.[145]

At about the same time, Buri[25] was also experimenting with small populations of *D. melanogaster*. These were segregating for two alleles at the brown eye-colour locus. The initial gene-frequency was 0.5 in every line, but after 10 generations the populations had drifted in both directions, so that all gene-frequencies were equally frequent. Populations were becoming fixed for one or other allele at equal rates. Again, it seemed that random processes were having an effect and directional selection was unimportant.

The laboratory environment is, however, different from nature, and all experiments that have involved the release of mutant *D. melanogaster* into natural populations seem to have ended in failure, or at least in the total disappearance of the mutant within a very few generations.[112]

STUDIES OF GENE-DISTRIBUTIONS IN NATURE

Industrial melanism

Despite the fact that many genes must have spread through a population at the expense of others, there are very few cases where this has happened and simultaneously been quantified by biologists. Many of the processes of evolutionary change, and especially neutral evolutionary change, are too slow to have been apparent within the time since the beginning of modern evolutionary genetics. There is, however, one notable exception: industrial melanism in the Lepidoptera. Here is a situation where genes have spread and been quantified within historical times. We shall discuss the phenomenon here with particular reference to the best worked example, *Biston betularia*.

Prior to 1848, this species seems only to have been known as the typical black and grey speckled animal, giving rise to its colloquial name of peppered moth. In that year the first all-black 'melanic' form was found in Manchester, and within 50 years this form had spread throughout urban Britain, increasing in frequency until it predominated in many areas. Its present-day distribution can be seen in Fig. 2.6.[148]

Laboratory experiments proved that the difference in pattern is due to a single locus, with the melanic gene being dominant. As we have already seen, the increase in the frequency of the melanic gene is far too rapid to be due to the pressure of mutation alone, and the association between high frequencies of melanics and industrial urban areas are too consistent to be due to chance, despite our ignorance of population sizes during the spread of the gene. Such consistencies between patterns of gene-distribution and environmental parameters are good *prima facia* evidence for selection, and so it has proved to be in *Biston betularia*. Haldane,[116] in a pioneering paper, estimated what the relative fitnesses must be if differential selection were responsible for the changes in gene-frequency. He suggested that in some urban localities, typical forms must have had a viability of only 0.66 relative to the melanics. We have seen that the probability of fixation of an allele is related to its selective advantage, and so it is perhaps not surprising that the melanic form of *B. betularia* is now to be found at a very high frequency in most urban centres of Western Europe.

Although in the 1920s the selective advantages had been estimated, it was still unclear what might be the selective agent. *B. betularia* is one of a large number of nocturnal moths that spend the day immobile upon exposed surfaces, and depend upon their personal crypsis for survival. The industrial revolution was accompanied by the production of a large amount of atmospheric pollution, which has continued in many areas to the present day. The toxic gases in the atmosphere killed many of the lichens and mosses that grew upon trees and walls. The associated particles of soot fell upon these surfaces and gradually turned them black. Upon these backgrounds, pale *B. betularia* are very conspicuous, and the mutant melanic form gained an immediate and enormous advantage.

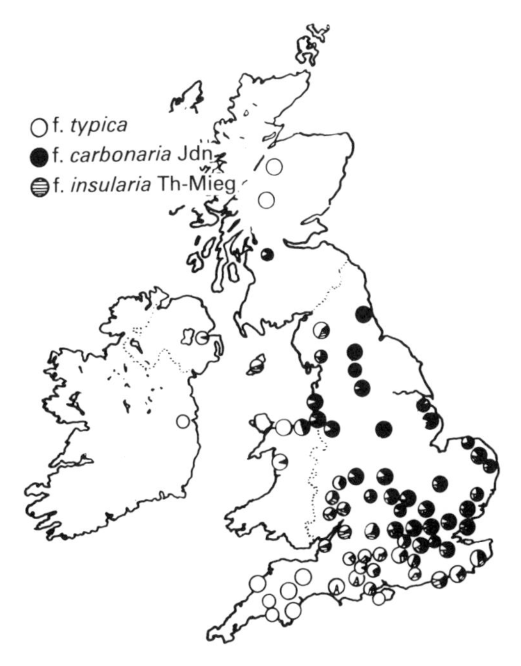

Fig. 2.6 The distribution of three phenotypes of *Biston betularia* in Britain. The melanic forms *carbonaria* and *insularia* occur in industrial locations and to the east, because the prevailing westerly winds carry pollution away from the sites of production. Data from Kettlewell.[148]

The crucial experiments proving that avian predators discriminate visually between the two phenotypes in urban surroundings were performed by Kettlewell.[148] He released marked male moths of both phenotypes in a park in urban Birmingham. Surviving moths were recaptured on subsequent nights using either a light trap, or a trap containing virgin females, which release a scent attractive to males. The numbers of moths released and recaptured in 1953 and 1955 are shown in Table 2.5, and it is clear that in both years a higher proportion of melanics were recaptured. This result could have been due to differential survival of the moths, but might equally have been due to different rates of recapture or emigration. Kettlewell eliminated such differences in behaviour by repeating the experiment in a rural area (Table 2.5). Here he found that a relatively higher proportion of typicals returned to the traps, showing fairly conclusively that the biological component which varied was the survival of the phenotypes.

Table 2.5 The results of Kettlewell's predation experiments involving the release and subsequent recapture of typical and melanic *Biston betularia* in industrial and rural environments. In all three experiments a statistically larger number of the form presumed to be cryptic were recaptured at mercury-vapour or female-assembly traps.

1953 Industrial Birmingham				
	Released	Returned	%	
Typicals	137	18	13·1	$\chi^2_{(1)} = 11.1$
Melanics	447	123	27·5	$P < 0{\cdot}001$
1955 Industrial Birmingham				
	Released	Returned	%	
Typicals	64	16	25·0	$\chi^2_{(1)} = 13.5$
Melanics	154	82	52·3	$P < 0{\cdot}001$
1955 Rural Dorset				
	Released	Returned	%	
Typicals	496	62	12·5	$\chi^2_{(1)} = 10{\cdot}0$
Melanics	473	30	6·3	$P < 0{\cdot}01$

He suggested that this differential survival was due to the activities of birds, and that the conspicuous moths were being selectively eaten. He established the truth of this suggestion by placing equal numbers of melanic and typical moths on trees, and watching the birds that came to eat them. Moths were replaced when they were eaten, and after a couple of days his results were quite conclusive. In the Birmingham park, 58 moths were eaten, of which 43 were typicals and 15 were melanics. In Dorset, however, out of 190 moths eaten, 26 were typicals and the remaining 164 were melanic. Taken in conjunction with the earlier results, this finding can leave little doubt that the birds, which were taking the visually conspicuous phenotypes, were responsible for the major part of the differential survival.

This research is classic, for it was the first well-established example of a connection between selective predation and large-scale changes of gene-frequency. The situation where a gene spreads through a population, replacing an alternative allele, is termed 'transient polymorphism' because it is essentially of a fleeting nature. In *Biston betularia*, the result has not been a total replacement of one allele by another, for in rural areas the typical form is still at an advantage. Even in the urban sites, typical moths still occur, but the typical gene is recessive, and we have already seen that it is difficult to totally remove such an allele. The last representative of a declining allele will usually occur in heterozygotes. Matings between two such heterozygotes will be very rare, but they will produce a few recessive homozygotes in each generation.

Industrial melanism is now a widespread phenomenon among the Lepidoptera, and its study has produced some very interesting results during the last few years. In 1964, the first legislation to restrict atmospheric pollution was passed in Britain, and certain local government authorities deemed it worthwhile to clean the filth and grime from the surfaces of their historic buildings. Cook *et al.*[56] have shown that in Manchester the frequencies of the non-melanic phenotypes in several species of moths have increased. Perhaps this is the beginning of the end for the industrial melanic.

The changes in gene-frequency that we have been discussing are large and rapid, so it is not altogether surprising that they are due to selection. However, other situations in other species might prove to be due to random processes. During the 1940s, attempts were made to study genetically controlled variation that was evident enough to be readily assayed, but was not so striking that selection would inevitably prove to be important.

Panaxia dominula

In their search for such material, Fisher and Ford[97] turned to another species of moth, but one which flies by day. The scarlet tiger moth, *Panaxia dominula*, is fairly widespread in Europe. At Cothill Fen, in England, there is a population that is polymorphic for the melanic gene called *medionigra*. This is not a very common gene even at Cothill, and apparently occurs naturally nowhere else in the world except as a very rare mutant. It is useful for the study of population genetics, because it shows incomplete dominance. The heterozygote can be readily distinguished from the two homozygotes. The 'wild type' or typical form has a black fore-wing with white spots, and a red hind-wing with black spots and patches. The heterozygote has fewer white spots on the fore-wing and the melanic has almost none. The black areas on the hind-wing are enlarged in the melanic, although the hind-wing is hidden from view at rest.

According to Ford,[102] the species possesses three means of defence. First, when the moth is at rest, the dappled white-on-black of the fore-wing is remarkably cryptic, especially when it sits in a situation where sunlight falls upon short grass and leaves. Second, when the animal flies, the flickering black of the fore-wings and red of the hind-wings create a disruptive pattern that makes it difficult to follow. Third, when a moth is disturbed by a potential predator, it exposes the red hind-wings and produces droplets of fluid from glands on the thorax. According to Fraser and Rothschild,[105–6] this fluid is distasteful to many potential predators, and the pattern of behaviour is probably to warn them of its unpalatability. Such warning behaviour is termed 'aposematic'.

In the years before 1930, it seems that the *medionigra* gene was rare at Cothill Fen. Very few specimens of *medionigra* moths are to be found in lepidopterist's cabinets, despite their predilection for varieties. Ford[102] estimates the frequency of the gene to have been no greater than 1.2%. In 1939, however, he found two *bimacula* and 27 *medionigra* in a random sample of 223 specimens, and the following year 24 out of 117. Thus, the frequency of the *medionigra* gene was 9.2% in 1939 and 11.1% in 1940. Here was a situation where the frequency of a melanic gene was changing in a population that was almost certainly free from visual predation.

Fisher[96] had shown that it is possible to predict the changes of gene-frequency to be expected from neutral alleles in a population of a known size. He and Ford began working with *Panaxia* to relate the changes in gene-frequency to the population size. It was a remarkable piece of work. They first derived an entirely new method for estimating population sizes and survival rates from the technique of mark and recapture, and then they applied it to *Panaxia* at Cothill Fen. In 1947, they reported the preliminary results, and the quantification of the population has continued to the present day.[102]

It soon became apparent that, while the changes in gene-frequency were not entirely consistent from year to year, there was a fairly steady decline in the frequency of the *medionigra* gene until about 1955. There followed a period of fluctuation around a frequency of 2% which has continued until the present day (see Fig. 2.7). These changes seemed too large to be due to drift, considering that the size of the population was never less than 1000 between 1941 and 1969. Selection, then, was probably involved in the changes in gene-frequency, though in their paper of 1947, Fisher and Ford were unable to say whether it was acting upon the locus which they were assaying.

The selective basis of this polymorphism was challenged by

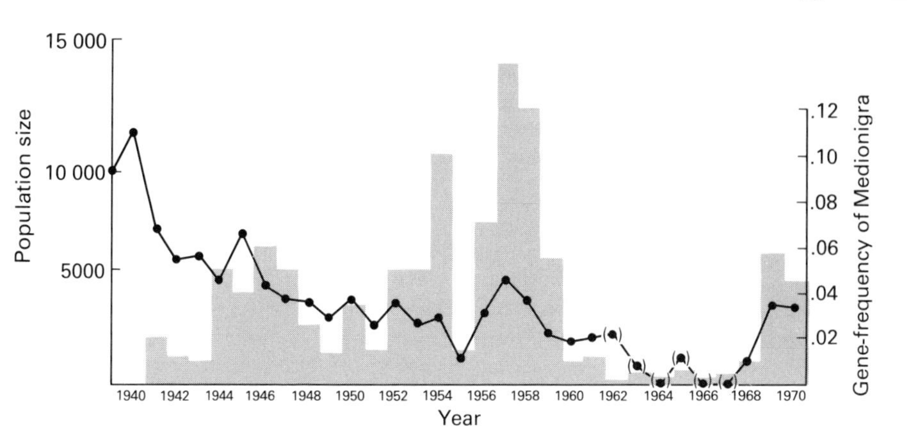

Fig. 2.7 The relationship between gene-frequency and population size in *Panaxia dominula* at Cothill Fen between 1939 and 1970. Gene frequency estimates based upon less than 100 individuals are shown in brackets. Data from Ford.[102]

Wright,[279] partly because of the inconsistency of the results, and partly on the grounds that the *effective* population size might be very much less than the actual number of flying adults. This, of course, is perfectly true, but a precise point upon which Wright criticized the results was that eggs might be laid in batches and random loss of whole families could then take place, which would reduce the effective population size. Sheppard[236] showed that the eggs are in fact scattered widely over the area by the female, and so families do not tend to be aggregated.

In this somewhat unsatisfactory state we must leave *Panaxia*. It seems fairly certain that selection is involved, though we do not know how. Non-random mating is a component of this selection (see Chapter 4) and a detailed analysis by Williamson[271] indicated that heterozygous advantage is not. Ford suggests[102] that there are other effects of the *medionigra* gene about which we know nothing, and this may well be true. The situation stands as strong evidence in favour of natural selection without the mechanism being understood.

Selection or drift in the land snail Cepaea nemoralis

More or less simultaneously with the studies of Fisher and Ford at Cothill Fen in England, Maxime Lamotte was assiduously travelling around post-war France collecting samples of the land snail *Cepaea nemoralis*. This species is highly variable in the patterns of colour and banding on its shell. Individuals may be brown, pink or yellow in colour, and may be banded (striped) or unbanded. The number of stripes, or bands, on the shell varies between one and five. The variation is mostly under simple genetic control. Breeding experiments[33] indicate that brown is dominant to pink, and pink is dominant to yellow, while the presence of bands is recessive to their absence.

It had been known for a long time that the proportions of brown, pink and yellow individuals vary from place to place, as does the frequency of banded and unbanded (see, for example, Diver[74]). It had been suggested by Diver that the variation was selectively unimportant, and that spatial changes in the frequencies of different phenotypes were the results of random fluctuations in gene-frequencies from generation to generation.

Among the hundreds of samples which Lamotte collected were 291 from populations whose size he had estimated from a mark and recapture technique. He divided these populations into those whose sizes were large, medium and small,[168] and compared the estimated frequency of the dominant gene for unbanded in the samples from the three classes. The results of this analysis are shown in Table 2.6. It is immediately apparent that most populations have rather few unbanded genes, and that very few populations have a lot. The variation between the gene-frequencies within a size class is given statistically by the variance. Lamotte showed that the variance is highest in the populations which he estimated to contain fewer than 500 individuals, and lowest in those with over 3000 individuals. He went on to propose that this resulted from the banded and unbanded genotypes being selectively neutral. He reasoned that, since the effects of random genetic drift are more important in smaller populations, these populations are more likely to diverge genetically one from another. We have seen this effect in the experiments of Kerr and Wright, and of Buri, discussed above. Before discussing the biological importance of Lamotte's findings, two facts must be mentioned. Firstly, there is no statistically significant evidence of differences between his variances. Secondly, his results may stem from the fact that smaller populations are more difficult to sample, and hence the estimates of gene-frequency will be less accurate than those derived from larger populations. This reduced accuracy may be reflected in a large variance which has no biological significance, being merely a statistical inevitability. Lamotte unfortunately gives no indication of his sample sizes, and so detailed re-analysis is not possible.

Accepting that his findings are real, however, leads to further trouble. The conclusions hinge upon the selective neutrality of the banded and unbanded genes, and indeed Lamotte's massive study

Table 2.6 The relationship between the estimated population size and the frequency of the unbanded phenotype in 291 populations of *Cepaea nemoralis* reported by Lamote.[168]

Population size	% Unbanded										
	0–10	10–20	20–30	30–40	40–50	50–60	60–70	70–80	80–90	90–100	Total
500–1000	32	22	14	10	7	8	6	4	2	4	109
1000–3000	33	22	17	8	7	8	6	0	4	0	105
3000–10000	32	15	8	8	7	4	1	1	1	0	77

gave no hard evidence of selection upon the manifestations of these alleles. However, at Oxford University, Cain and Sheppard[31] accumulated a series of samples of *C. nemoralis* from the regions around Oxford, taking care to collect each sample from a small area, to ensure that all individuals came from the same random-mating population.

They recorded the frequency of yellow snails in each sample, and considered shells to be 'effectively banded' if stripes were present on their upper surface, or 'effectively unbanded' if they were not. Their first results[31] indicated that samples from the same habitat resembled each other in phenotype frequency rather more closely than would be expected by chance. This result was confirmed in a more extensive survey[32] and Fig. 2.8, taken from that study, shows a clear separation of woodland and non-woodland samples. On the whole, woods contain fewer yellow shells and more 'effectively unbanded', while

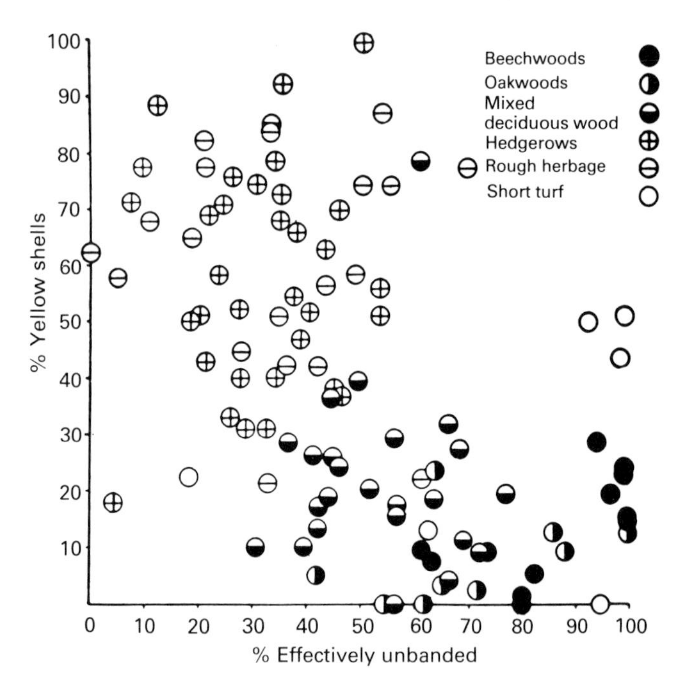

Fig. 2.8 Scatter diagram showing the relationship between colour and banding in a series of samples of *Cepaea nemoralis* from the area around Oxford in England. Note the differences in phenotype composition of the samples from different habitats. Data from Cain and Sheppard.[32]

non-woodlands have higher proportions of yellow shells and fewer 'effectively unbanded'.

Consistent results like this usually have some underlying biological basis. We shall see that natural selection for crypsis is a major factor influencing phenotype frequencies in Oxfordshire, but we need to know why Cain and Sheppard[31–2] obtained results at variance with those of Lamotte, who considered his data to indicate selective neutrality. Lamotte[168] also studied the movement of *C. nemoralis*, by marking snails with spots of paint. He estimated that no population larger than sixty metres in diameter could be regarded as random mating. Linear colonies, such as roadside verges or hedgerows, greater than about 50 metres in length were similarly too large for random mating to be likely. By his own admission, some of Lamotte's 'populations' were linear and were over 150 metres in length. Others were several thousand square metres in area. Sheppard[238] showed that a single widespread colony near Oxford occupied an area of downland turf, which included three isolated beechwoods. The samples from these woods were very different in phenotype composition from downland samples taken only a few metres away.

Lamotte gives no information concerning the uniformity or otherwise of the habitats from which his samples come. Consequently, it is difficult to comment upon the possibility that he unconsciously lumped heterogeneous data. The possibility remains, however, that his larger populations occupied several different microhabitats. If this were the case, these samples would reflect the average composition of several subpopulations or demes. They would then have a lower overall variance than samples from smaller populations which reflect the situation in individual habitats.

Lamotte's results have often been quoted as evidence of neutral genes. Regrettably, in the absence of detailed information concerning the techniques of sampling and analysis, we must dismiss it as 'not proven'. Furthermore, in addition to the circumstantial evidence provided by Cain and Sheppard,[31–2] *direct* evidence is now accumulating that *C. nemoralis* is subject to visual selection for crypsis.

It is well known to European naturalists that song thrushes (*Turdus ericetorum*) and many other predators feed upon land snails. Predators which hunt for the snails by sight would be expected to take a relatively larger proportion of visually conspicuous phenotypes, just as they do with *Biston betularia*. On shady woodland floors, especially upon a carpet of dead leaf litter, yellow snails appear to be conspicuous, and ought to be selectively eaten by predators. In grassland, however, the yellow-shelled snails (which look greenish in life) should be better camouflaged. In a beechwood, the ground layer is much more uniformly brown than in oak or sycamore wood. Unbanded snails might be better camouflaged in the former and banded in the latter. These patterns of selection have been postulated by Cain and Sheppard[31–2] to account for the distribution of phenotypes that they observed. Experimental support for these postulates is still somewhat scanty, although the material is ideal for field study. The song thrust is a principal predator of *Cepaea*, and has the experimentally convenient habit of taking its snails to a central stone (or 'anvil') where it smashes them—leaving the shell behind for the experimenter to collect at his leisure.

Sheppard[237] released a sample of *C. nemoralis* marked with paint spots in a woodland near Oxford. He showed that the proportion of yellow snails killed by the thrushes declined during the period from April to June in 1951. He suggested that this was due to the selective advantages changing as the summer vegetation matured, and the background became green. This is probably correct, although the possibility of changes in the preferences of the thrushes, or even in the thrushes themselves, cannot be ruled out. A related series of experiments was reported by Arnold.[6] He also released large samples of snails, marked with scratches on the shell, in a variety of woodland and grassland sites in south-eastern England. In one series of experiments he offered the predators a choice between two phenotypes, and found that the visually conspicuous form was taken in excess in all his woodland localities. The situation in grassland was rather more confused and warrants further study.

Studies of the ecological genetics of other species of molluscs have been undertaken by a variety of workers on a variety of species. All of them have suggested that natural selection has an important influence on the frequencies of shell-character phenotypes. Clarke[38] showed that the patterns of these distributions in *Cepaea hortensis* are consonant with visual selection for crypsis. Parkin[209] has shown the same for *Arianta arbustorum*, and has provided evidence (from broken shells) that predators do in fact take excesses of the visually conspicuous phenotypes.

The selective importance of characters which affect the appearance of an organism is almost inevitable. When a pair of alleles produces alternative phenotypes that are sufficiently distinct for them to be separated visually, it is extremely probable that there will be a corresponding difference in some aspect of their biology. These differences need not necessarily involve predator-prey interactions. We shall see later that the appearance of an animal may easily affect its thermal properties, and hence its behaviour. In plants, variation in flower colour can have important significance in reproduction, through the colour preferences and behaviour of pollinating insects. Evidence of natural selection upon such major differences in phenotype is to be expected, and the pioneering studies are so well-substantiated that additional examples excite little comment from contemporary ecological geneticists.

3

Molecular Evolution

INTRODUCTION

Perhaps the major development in biological science during the 1960s was the explosion in research into molecular biology. New developments and discoveries in this field caused a revolution in our thinking about the structure and function of the genetic material. The advances made in molecular genetics have had their influence upon our understanding of the evolutionary process in one particularly important way, for it is now possible to approach closer than ever before to the level of the individual gene. We shall see, however, that the search for evidence relating to the nature of evolution at the molecular level is still fraught with unresolved difficulties centred especially around the relative importance of natural selection and random processes.

In 1955, Sanger published the first complete amino-acid sequence of a complex biological molecule.[232] He studied insulin because it is a fairly small protein, consisting of only 51 amino acids, but even so it took him almost six years to solve the biochemical problems of sequencing such a large organic molecule. He found that it consisted of two chains that could fairly easily be separated. The A-chain comprises 21 amino acids and the B-chain contains 30. He used digestive enzymes to chop these two sub-units up into smaller pieces and then proceeded to identify the amino-acid composition of each small piece. He gradually produced a large number of bits of information, each of which gave the sequence of a few adjacent amino acids. There was often some overlap between the short sequences, and so he gradually built up a single unique map of each of the two chains of the insulin molecule.

Since this brilliant pioneering research, many molecular biologists have applied themselves to protein sequencing, and the techniques have been refined and developed until the time involved has been reduced from a matter of years to one of months. New molecules have been examined, and the same molecules from a variety of species, until now sequence data are produced so rapidly that it is necessary to publish atlases of protein structure so that other workers can keep track of the results. The best and most widely available is that edited by Margaret Dayhoff[71] which is updated regularly.

The data that these labours have produced are of unique value to the evolutionist. The amino-acid sequence of a polypeptide chain is related, via the genetic code, to the sequence of bases in the DNA molecule; i.e. to the very gene itself. By comparing functionally related polypeptide sequences, we can begin to see how the genetic material within a species has changed and evolved during its history. Furthermore, by comparing the same protein in a variety of species, we can examine the similarities and differences between genes of those species.

This kind of research has only been possible since about 1966. It has given rise to the new and exciting science of molecular evolution. In the following pages we will see how molecular evolutionists have cast light upon the evolution of those globins that are found to be present in man. We will also see how evolutionary trees can be built by comparing and contrasting the amino-acid sequence of the widespread organic molecule cytochrome-c. In passing, we will discuss the problems and arguments that have beset the new science, and show how its devotees have uncovered new evidence on the relative importance of selection and neutrality in the evolutionary process.

EVOLUTION OF HUMAN GLOBINS

We will start by considering how the sequence data accumulated by armies of molecular biologists can be used to disentangle the evolution of a 'family' of proteins within a single species. The best example stems from studies of the structure of the various types of oxygen-bearing globin that are found in our own species. These researches into the amino-acid sequence and the three-dimensional structure of the haemoglobins have cast light upon evolutionary changes in the very composition of genes.

Haemoglobin is a tetramer: a complex molecule consisting primarily of four long chains of amino acids. These chains are joined together at one end, and X-ray crystallography has shown that they are coiled in a complex fashion. This coiling probably stems from the primary structure (i.e. the sequence in which the amino acids occur in the polypeptide) and typically forms four 'pockets' in the haemoglobin molecule. Each pocket contains a prosthetic haem group, a unit composed not of amino acids, but of a porphyrin ring with an iron atom in the centre. Oxygen binds to the iron atom in each group, but can easily be released at the site where it is needed for respiration. Haemoglobin is thus essential as a respiratory pigment in man and most other vertebrates.

Several different forms of haemoglobin are known to occur in man, the most abundant being haemoglobin A (HbA). The sequence of amino acids in HbA was determined by Braunitzer *et al.*,[16] who showed that two of the four polypeptides contained 141 amino acids, and the other two consisted of 146. The shorter chain is called the α polypeptide, and the other is called the β polypeptide. HbA can thus be written $\alpha_2\beta_2$. Part of the sequence of the α and β polypeptide chains is given in Table 3.1, where it is readily apparent that there is a considerable similarity between them. There are problems in aligning the sequences precisely, but a reasonable arrangement can be made by leaving gaps in appropriate places, for example at site 2 in the α chain.

The amino acid sequence of a polypeptide is related directly to the sequence of bases in the DNA molecule via the genetic code. The similarity in the composition of the α and β haemoglobin chains thus probably reflects a very close homology in the base sequences of the genes that code for them.

There are two other haemoglobins that occur widely in human blood. The first of these is found primarily in the foetus, and forms 70–80% of haemoglobin in the newborn. This is also a tetramer, but does not contain the β polypeptide. The chain that is present is called δ, and this has also been sequenced (Table 3.1). Like the β polypeptide, it is composed of 146 amino acids, and the two sequences are very similar, having the same amino acid at 107 sites.

The other haemoglobin that is widely distributed in human populations is called A_2. It is normally present as about 2.5% of the total adult haemoglobin, and again does not contain the β polypeptide. This is replaced by γ haemoglobin chain, that resembles β and δ in consisting of 146 amino acids. Its resemblance to β is even more profound, there being only 10 amino-acid differences in the entire length of the polypeptide.

It is apparent that all these sub-units are extremely similar in their amino acid sequences and this must surely reflect a close resemblance

Table 3.1 The first 40 amino acids in the sequences of α, β, γ, and δ haemoglobins from man, and of human myoglobin. Note that the judicious insertion of blanks into the sequences allows 9 and possibly 10 of the amino acids to be identical in all five polypeptides. The blanks would correspond to positions where a complete codon has been deleted from the DNA molecule.

	1	2	3	4	5	6	7	8	9	10	11	12	13	14	15	16	17	18	19	20
α	VAL	—	LEU	SER	PRO	ALA	ASP	LYS	THR	ASN	VAL	LYS	ALA	ALA	TRY	GLY	LYS	VAL	GLY	ALA
β	VAL	HIS	LEU	THR	PRO	GLA	GLA	LYS	SER	ALA	VAL	THR	ALA	LEU	TRY	GLY	LYS	VAL	—	—
γ	GLY	HIS	PHE	THR	GLA	GLA	ASP	LYS	ALA	THR	ISO	THR	SER	LEU	TRY	GLY	LYS	VAL	—	—
δ	VAL	HIS	LEU	THR	PRO	GLA	GLA	LYS	THR	ALA	VAL	ASN	ALA	LEU	TRY	GLY	LYS	VAL	—	—
MYO	GLY	—	LEU	SER	ASX	GLY	GLX	TRY	GLU	—	VAL	LEU	ASX	VAL	TRY	GLY	LYS	VAL	GLA	PRO
															*	*	*	*		

	21	22	23	24	25	26	27	28	29	30	31	32	33	34	35	36	37	38	39	40
α	HIS	ALA	GLY	GLA	TYR	GLY	ALA	GLA	ALA	LEU	GLA	ARG	MET	PHE	LEU	SER	PHE	PRO	THR	THR
β	ASN	VAL	ASP	GLA	VAL	GLY	GLY	GLA	ALA	LEU	GLY	ARG	LEU	LEU	VAL	VAL	TYR	PRO	TRY	THR
γ	ASN	VAL	GLA	ASP	ALA	GLY	GLY	GLA	THR	LEU	GLY	ARG	LEU	LEU	VAL	VAL	TYR	PRO	TRY	THR
δ	ASN	VAL	ASP	ALA	VAL	GLY	GLY	GLA	ALA	LEU	GLY	ARG	LEU	LEU	VAL	VAL	TYR	PRO	TRY	THR
MYO	ASX	ISO	ALA	GLY	HIS	GLY	GLX	GLX	VAL	LEU	ISO	ARG	LEU	PHE	LYS	GLY	HIS	PRO	GLA	THR
						*		*?		*		*						*		*

Amino acid code:

ALAnine
ARGinine
ASN = Asparagine or Aspartic acid
ASPartic acid
ASX = Asparagine
CYSteine·
GLA = Glutamic acid
GLUtamine
GLX = Glutamine or Glutamic acid
GLYceine
HIStidine
ISOleucine
LEUcine
LYSine
METhionine
PHEnylalanine
PROline
SERine
THReonine
TRYptophan
TYRosine
VALine

in the base sequence of the genes concerned with their production. This resemblance could be due to chance, or natural selection might have produced the several haemoglobin polypeptides separately at a series of different genetic sites. However, the probability of such homology being produced in a series of functionally related molecules at *independent* loci must be vanishingly remote even by natural selection. A much more probable explanation is that all the genes concerned have evolved from one ancestral 'globin' gene by a series of duplications followed by varying amounts of divergence. We will briefly follow this argument to see where it leads.

The three haemoglobins that we have discussed are all tetrameric. There is, however, a monomeric globin, called myoglobin, that is also present in man. This molecule is found commonly in muscle, and it also has the potential to combine reversibly with oxygen. Its function is to store oxygen and to transmit it from the red blood cell to the precise site where it is to be used. Its amino-acid sequence is again similar to haemoglobin,[268] although its similarity is less spectacular than the chains of tetrameric haemoglobin. Nevertheless, this similarity must be too great to be due to chance.

Ingram[129] discusses the possible evolutionary history of all these globins in some detail. It seems intuitively likely that monomeric globins are the most primitive, and that the early development concerned one gene that coded for a single oxygen-bearing polypeptide. At some stage during the history of this gene, a duplication occurred, and so the same molecule was coded at two different sites. Specialization occurred alongside divergence of the two genes, one to give rise to present-day myoglobin and the other to produce the α chain gene. At some stage, modifications arose to produce a dimeric, and subsequently a tetrameric, α molecule. Ingram suggests that the evolutionary advantages of these developments would be considerable, since they might entail the possibility of interactions between the haem groups and a consequent increased efficiency in the uptake and release of oxygen.

During the evolution of the lower chordates, a second duplication must have occurred, this time at the α locus. There were now two loci producing an identical α chain for incorporation into the α_4 molecule, and the opportunity for specialization and sequence divergence arose again. One gene evolved towards the present day β–γ sequence, and the other further towards contemporary α. This resulted in the evolution of the characteristic molecule consisting of a pair of α chains and a pair of some other haemoglobin chains.

The situation at this stage of man's evolutionary history was that a monomeric myoglobin molecule occurred alongside a tetrameric 'archae-haemoglobin'. These were coded by three different loci, and the near-ubiquitous presence of the α chain in contemporary haemoglobins suggests that all subsequent evolution must have involved the non-α locus. Further duplications occurred at this site to produce the β, γ and δ genes which resemble one another very closely. A sketch of this evolutionary tree is given in Fig. 3.1.

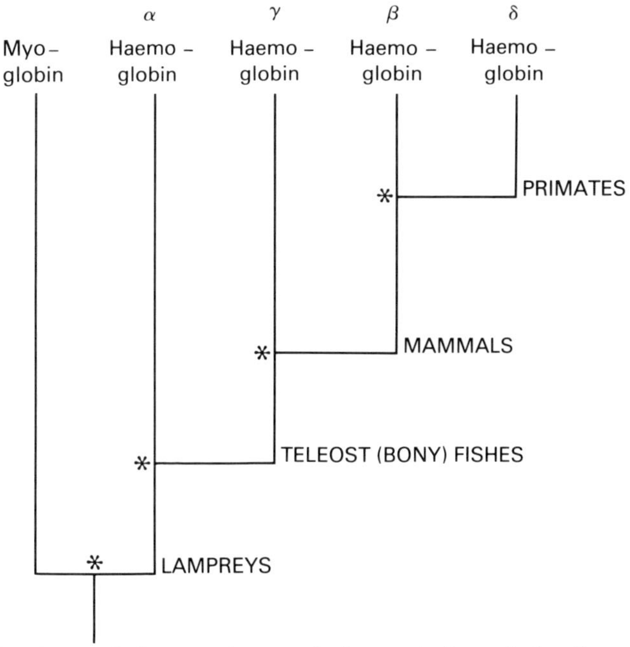

Fig. 3.1 An evolutionary picture of the evolution of the five globin polypeptides to be found commonly in man. The asterisks indicate points at which a gene duplication is postulated, and the groups of animals in which the particular duplication is to be found are listed to the right.

There is supporting evidence for this model of globin evolution from studies of comparative anatomy and physiology. Myoglobin or other monomeric globins are present in organisms as diverse as *Chironomus thummi* (a non-biting midge), *Aplysa limacina* (a gastropod mollusc) and man. The sequences are surprisingly similar. For example, the haem-bearing globin of *C. thummi* contains 136 amino acids (only 16 less than human myoglobin), and the polypeptides are similar at about 15% of the sites.[71] If the arrangement were random, we would expect concordance at only 1 in 20 of the residues—about 5%. This similarity suggests a common ancestry of the gene and indicates strongly that the globin locus is older than the mollusc and arthropod phyla.

Tetrameric haemoglobins are found among all the vertebrates from the teleosts 'upwards', but a monomeric molecule is found alone in the lamprey. We can presume, then, that the critical evolutionary steps of duplication, dimerization and tetramerization took place during the early evolution of the vertebrates. This is confirmed by the discovery of sequences clearly recognizable as α from the carp and β from the frog.[71] The δ polypeptide has only been found in the primates, and may be restricted to this group. The γ chain seems only to have been sequenced in man, so its evolutionary status is unclear. However, we can use these facts to incorporate a time scale into Fig. 3.1, and thereby produce a remarkable picture of the evolutionary history of a series of different loci within a single organism.

These human globins have given us an insight into the way in which data from molecular biology, when supported by information from more classical studies, can be used to disentangle the evolutionary history of a species. Sequence data can do more than this, however. The next section will discuss how data on the amino-acid sequence of the *same* protein from a variety of species can be used to provide an assessment of their evolutionary relationships. The resemblance between these results and those of 'classical' systematics is so profound as to give rise to intense satisfaction on both sides.

SOME EFFECTS OF MUTATION ON AMINO-ACID SEQUENCES

It has already been suggested that the amino-acid sequence of a polypeptide is related to the structure of the DNA molecule through the genetic code. Differences between two species in the structure of a particular protein reflect the differences in the base sequence of the gene that produces it. Sequence analysis has shown repeatedly that functionally similar proteins often have a basically similar structure in a wide range of species. The implication of this is that the species had a common origin, and that the differences between the sequences reflect differences within the evolutionary history of the species concerned.

If representatives of a particular protein from two separate species differ at a single site, we can infer that some mutational event occurred in one or other of the lineages to change the ancestral gene at one codon.

There are several ways in which mutation can effect the amino-acid sequence of a protein. The simplest of these is direct substitution of one base for another. For example, the genetic code (Table 3.2) shows us that the codon GTT codes for glutamine; if either glycine or adenine is substituted for the terminal thymine, the triplet will code for histidine. The polypeptide will differ at one, and only one, of its constituent amino acids, so there has been, effectively, a 'point mutation'.

Table 3.2 The genetic code in DNA. The code is read in triplets of bases, and each triplet codes for a single amino acid. Those triplets marked with a star are the so-called nonsense codons, or punctuation marks.

		3rd base			
1st base	2nd base	A	G	T	C
A	A	PHE	PHE	LEU	LEU
	G	SER	SER	SER	SER
	T	TYR	TYR	*	*
	C	CYS	CYS	*	TRY
G	A	LEU	LEU	LEU	LEU
	G	PRO	PRO	PRO	PRO
	T	HIS	HIS	GLN	GLN
	C	ARG	ARG	ARG	ARG
T	A	ILEU	ILEU	ILEU	MET
	G	THR	THR	THR	THR
	T	ASN	ASN	LYS	LYS
	C	SER	SER	ARG	ARG
C	A	VAL	VAL	VAL	VAL
	G	ALA	ALA	ALA	ALA
	T	ASP	ASP	GLU	GLU
	C	GLY	GLY	GLY	GLY

It is important to notice here that the genetic code is redundant: that is, several triplets code for the same amino acid. For example, the codons GTT and GTC both code for glutamine, and they are said to be 'synonymous'. Substitutions of bases that result in an interchange between two synonymous codons are not, of course, detectable by amino-acid sequencing, and about 20% of the possible base substitutions are of this kind.

We should also note here that certain amino-acid substitutions need two base changes. For example, tryptophan is coded by the triplet ACC. Glutamine (GTC or GTT) can only be converted to trytophan if two or three base substitutions have taken place. Analysts of base sequences conventionally take the minimum number of mutations required to replace one amino acid for another in a polypeptide.

Finally, we must consider the effects of adding or removing bases from the DNA molecule. Because the bases are read in groups of three, the addition or deletion of one or two bases will usually result in the complete disruption of the polypeptide from the point of mutation. The addition or deletion of *three* bases will, of course, result in the addition or deletion of one amino acid. The remainder of the polypeptide will remain unchanged.

THE CONSTRUCTION OF EVOLUTIONARY TREES USING AMINO-ACID SEQUENCES

We can now begin to see how sequence data can be used to disentangle the evolutionary relationships between species, by considering the structure of α-haemoglobin from the chimpanzee and the gorilla. In both species, the α-haemoglobin molecule consists of a chain of 141 amino acids, and the sequences are identical, except for site 23 where glutamine occurs in the chimpanzee and asparagine in the gorilla. From that information we can say that, following the separation of the ancestral stocks of these two species, a mutation occurred and spread. Without further information we cannot say whether glutamine became substituted for asparagine in the chimpanzee line, or asparagine for glutamine in the gorilla line. When we learn that man, rhesus monkeys and *Irus* maraques all have glutamine at site 23, it becomes more probable that it is the gorilla which carries the gene resulting from a recent mutation. It is still possible that the same mutation has occurred in all primates apart from the gorilla, but this is more unlikely.

There is, of course, a third possibility. The ancestral primate stock may have carried neither asparagine nor glutamine at site 23, and substitutions may have taken place in both gorillas and the rest. This again is less likely and in the absence of evidence to the contrary, analysts of sequence data assume that the minimum number of amino-acid substitutions have occurred, when they derive relationships between the proteins of different species.

One of the most comprehensive analyses of the evolution of a protein concerns cytochrome-c which is a key enzyme in oxidation reactions and seems to occur in practically every living organism. The sequence data for cytochrome-c was first analysed by Fitch and Margoliash[98] and more recently by Dayhoff *et. al.*[72] The former based their analysis upon 20 species, the latter had the benefit of several further years of sequence analysis and were able to use 33.

Fitch and Margoliash used a method that simply counted the number of amino-acid sites at which two species differed, and corrected any differences in chain length by relating them to an arbitrary chain length of 110 amino acids. For example, *Saccharomyces* and *Candida* have 109 comparable amino-acid sites in their cytochrome-c sequences, so the number of sites at which they differ is multiplied by 110/109. The number of differences is then rounded to the nearest whole number and the construction of an evolutionary tree can begin.

A simple situation is one where there are three species X, Y and Z whose sequences differ as shown in Fig. 3.2. Clearly X and Y are the closest, since they differ from one another at only 10 sites. We can therefore produce an initial family tree as in the lower left part of Fig. 3.2. The number of substitutions that have taken place in each lineage since the separation of species X and Y are represented by a and b respectively. The number of substitutions which separate the common ancestor of these species from the present day sequence of species Z is represented by c.

X – Y differ at 10 sites

X – Z differ at 17 sites

Y – Z differ at 19 sites

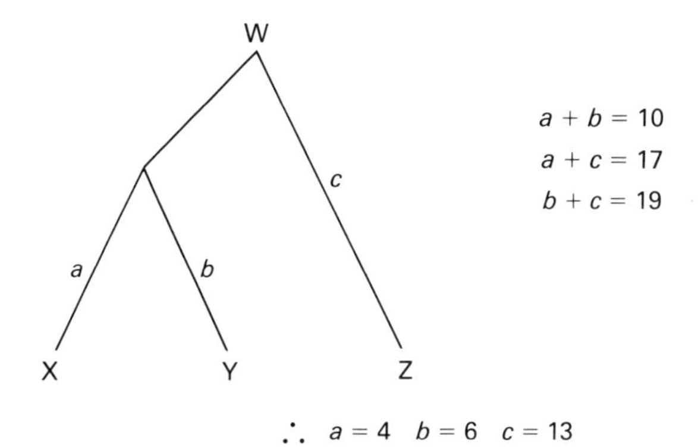

Fig. 3.2 A hypothetical relationship between three species which differ at various numbers of sites in their amino-acid sequence. See text for details.

It is fairly straightforward to solve the values of a, b and c. The value of $a+b$ is equal to the number of substitutions separating the species X from species Y. Similarly, $a+c$ and $b+c$ refer to the number of substitutions by which X and Z, and Y and Z differ. These three simultaneous equations are solved to give the values of *a*, *b* and *c* shown in the bottom right-hand part of Fig. 3.2. From this we can deduce that, since their separation, four substitutions have taken place in the line leading to species X, and six in that leading to species Y. Thirteen substitutions separate their ancestor from the third species, Z. We cannot infer the position of W, the common ancestor of all three species, from these results, nor can we tell precisely *which* substitutions are involved in the various branches of the evolutionary tree.

When more than three species are involved, this exercise rapidly becomes very complex. The two species with the most similar sequences are paired as X and Y, and compared in turn with each of the others. Mean values for *a*, *b* and *c* are produced, and the next nearest species is taken and compared with these two. Gradually the picture becomes more and more complete, until all the species for which sequences are known have been incorporated, and the mean number of substitutions between each species group and its common ancestor has been estimated.

Unfortunately, this method neither tells us *which* substitutions are involved in a particular evolutionary pathway nor the *order* in which they have taken place. The principles are more simple than those of Fitch and Margoliash but more time-consuming when there is a large amount of data.

The sequences are first written down underneath one another, and then shuffled to-and-fro until the best possible match between one species and the next is obtained. This is fairly easy for cytochrome-c, because, of the 104–112 amino acids composing the molecule, no less than 35 appear to be in homologous positions in all the species that have been sequenced. With some proteins it is necessary to leave a space to maintain the regularity of the constant sites, and to explain this as the result of an insertion or deletion of an amino acid but this is not necessary for cytochrome-c. It is apparent that there are also amino acids missing at the ends of the chain in some vertebrate species (see Table 3.3).

The derivation of an evolutionary tree from these matched sequences is explained by Dayhoff *et al.*[72] using sites 8, 15 and 104 from cytochrome-c as examples, and comparing only four species. The amino-acid present at these sites in man, sunflower, fruit fly and baker's yeast are shown in Table 3.3, from which it will be seen there is no amino acid at all at site 8 in man. The principles of the method become apparent if we consider the sites in turn.

Site 8. There are two alternative hypotheses here. Firstly, the ancestral sequence had no amino acid at this site, and flies, sunflowers

Table 3.3 The amino acid found to be present at three chosen sites in the cytochrome-c molecules from four widely different species.

Organism	Site 8	Site 15	Site 104
Baker's yeast	alanine	alanine	threonine
Sunflower	alanine	alanine	alanine
Fruit fly	alanine	lysine	alanine
Man	missing	lysine	alanine

and baker's yeast have all had an amino acid inserted following the separation of man from the main stock. Secondly, man suffered a deletion following his separation as a species. We cannot rigorously distinguish between these hypotheses, but *a priori* it is more likely that one species has lost an amino acid that many gave gained one.

Site 15. The fact that men and fruit-flies carry one amino-acid at this site, and that sunflowers and baker's yeast have another may imply a closer relationship of the former to each other than to the latter (which might also form a related pair). We cannot say whether lysine was substituted for alanine in the line leading to man and the fruit fly, or whether the reverse happened in the sunflower-baker's yeast stock. Indeed, both amino acids might be the result of substitutions for a third amino acid in the ancestral species.

Site 104. The situation here is similar to site 8, except that there is a substitution of threonine for alanine in yeast, whereas site 8 showed a loss of alanine in man. The same argument applies, however, and we postulate the substitution of threonine after baker's yeast separated as a species in its own right.

A simple evolutionary tree of the cytochrome-c molecule can now be constructed, based upon these three sites (Fig. 3.3) (see page 62).

As with the method of Fitch and Margoliash[98] the analysis obviously becomes very complex as the number of sites and species increase. Dayhoff *et al.*[72] used a computer to shorten the time needed to construct the best evolutionary tree, always using the rule of reducing the number of substitutions, deletions or insertions to the minimum. Their resulting evolutionary tree is shown in Fig. 3.4 (see page 63. They estimate the closeness of relationship by the number of substitutions (per 100 sites) necessary to change the cytochrome-c of one species into another. It is possible to produce sequences for the nodes in the evolutionary tree (represented by numbers in Fig. 3.4). Thus, the exact processes of evolutionary change can be observed in this model, which represents its great advantage over earlier ones depending only upon the number of substitutions.

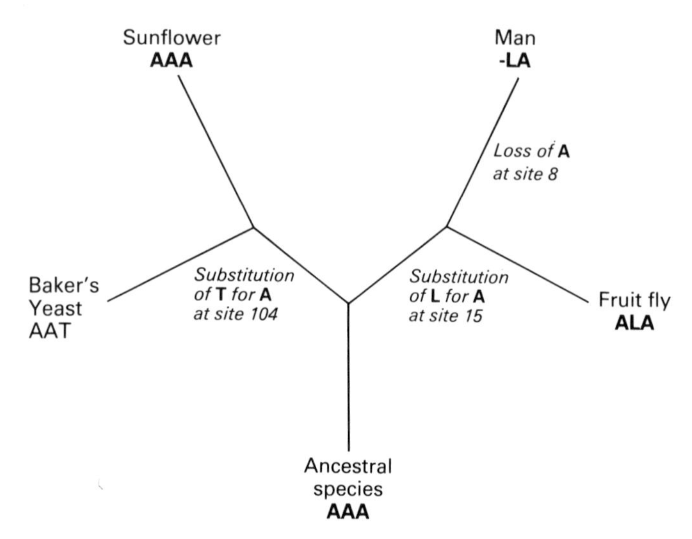

Fig. 3.3 The evolutionary relationships of the four species listed in Table 3.3, based upon the amino acid to be found at sites 8, 15 and 104 of the cytochrome-c molecule. The ancestral species may have had alanine *or* lysine at site 15, the data from Table 3.3 is insufficient to allow a precise determination. We have assumed here that the ancestral sequence was AAA.

In their pioneering study, Fitch and Margoliash found that they had one or two anomalous results. For example, they found that the kangaroo was rather closer to the non-primate mammals and more distant from the primates than is expected from studies of anatomy and palaeontology. Conventional systematics tells us that marsupials should be more distant from higher mammals than these are from each other. This discrepancy is reduced somewhat in the later study, and perhaps as more species become available, it will become even less. It would be interesting to have more sequences from lower vertebrates to examine the relationships between the amphibia and reptiles, and also to see how lungfish and coelacanths fit into the scheme.

One great advantage of a family tree derived in this way is its objectivity; a few simple rules, such as that requiring the minimum of substitutions, and the computer can do the rest. Much of classical systematics uses the taxonomist's judgement, which is perfectly acceptable, but nevertheless introduces a subjective element. The recently developed techniques of numerical taxonomy[248] have done much to remove this subjectivity, but the option of including or

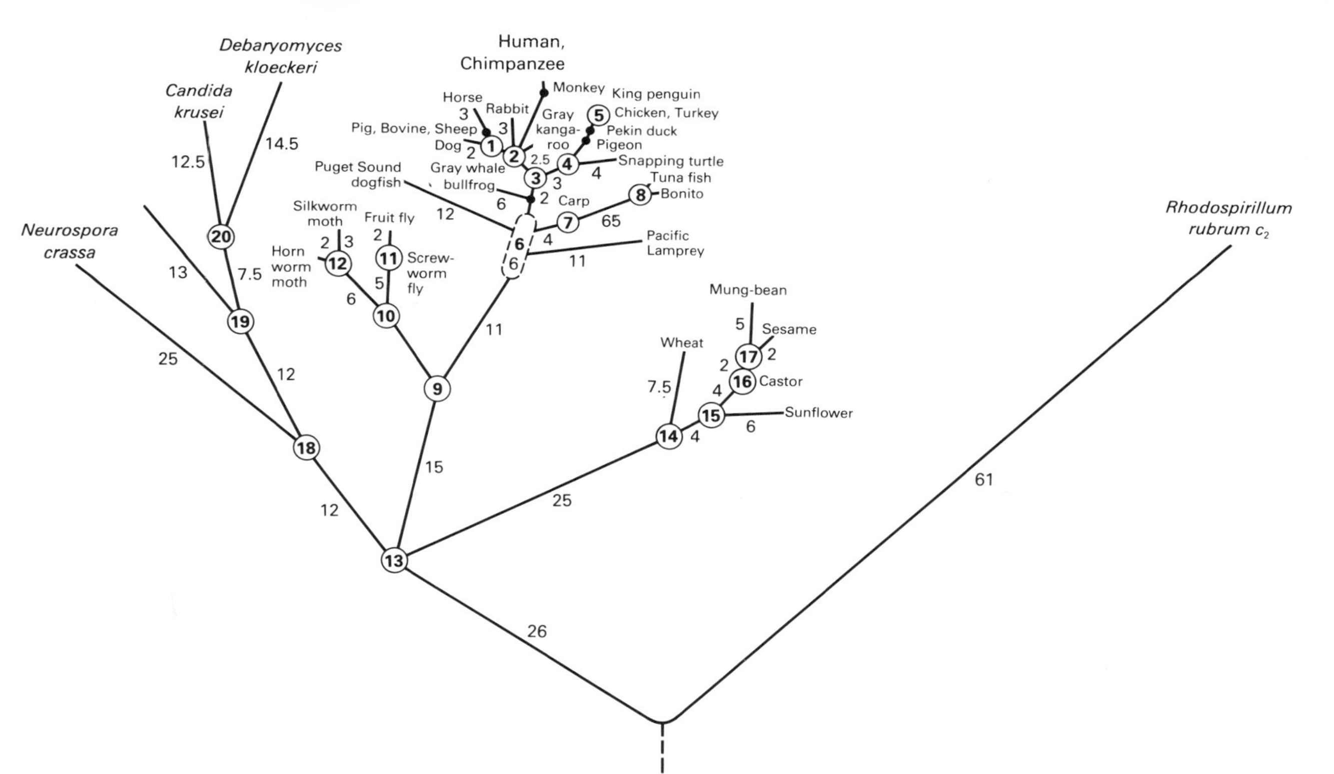

Fig. 3.4 An evolutionary tree of a wide variety of organisms based solely upon the amino-acid sequences of cytochrome-c From Dayhoff *et al.*[72]

omitting characters still remains. An evolutionary tree built from protein sequences does not suffer from this difficulty (unless, of course, gaps in the sequence are filled in by referring to the atlas!). It will be interesting to see how well evolutionary trees constructed from proteins other than cytochrome-c resemble the traditional pattern. The sequence systematists must take great encouragement from the similarity of their evolutionary trees with more traditionally derived results. So far, it has not proved necessary to consider the need for revision of any taxonomic category as the result of sequence analysis. If and when this becomes a matter of dispute between the classical and the newer approaches to systematics, there will be considerable interest to the bystander to observe the arguments and outcome.

THE RATE OF EVOLUTION

A valuable aspect of the analysis of protein sequences is that it allows us to estimate the *rate* at which evolution has taken place. We have already considered the probability of the substitution of a newly arisen mutation under the twin hypotheses of selection and neutrality. It is possible to turn to the question of the speed with which such substitutions have occurred for individual protein-forming genes during evolution. Armed with this information, it might be possible to determine whether they have been predominantly under the influence of selection, or whether random processes have played a major role, although we shall see that no clear-cut decision has yet emerged.

In a fundamental but rather difficult paper, Haldane[120] calculated the relationships between the strength of selection, the amount of mortality due to this selection (which he called the 'cost of evolving'), the size of the population, and the number of generations needed for the substitution of one allele by another. He concluded that, provided selection is not too intense, the number of deaths needed for the substitution of one allele by another would be virtually independent of the intensity of selection. He considered that it should equal about thirty times the population size. If 10% of the individuals in a population died as the result of selection at a locus, he estimated that about 300 generations would be required for the substitution of a new allele.

Zuckerkandl and Pauling[283] first applied the data from protein sequences to this problem. They argued that mutational changes within a gene are likely to occur at random, and that the probability of the successful substitution of one allele by another is slight. As a consequence, they reasoned that amino-acid substitutions themselves can be regarded as independent both in type and position. The pattern of successful substitutions, or, to be more precise, the pattern of those substitutions which we observe or infer from sequence data, is likely to approximate quite closely to a random or Poisson distribution. The frequency of unchanged sites during a given time interval is thus e^{-k}, where k is the mean number of substitutions per site during that time interval. Hence the proportion of amino-acid sites where at least one substitution has occurred is $(1-e^{-k})$. Zuckerkandl and Pauling estimated the value of this from the number of invariant sites in the sequences of proteins of the same length in two species, as follows:

$$\text{number of sites at which the amino acids differ} = (1-e^{-k}) \times \text{total number of sites}$$

Kimura and Ohta[154] demonstrated the use of this formula with sequence data from the α chain haemoglobin of man and the carp. Comparisons show that these two proteins have the same amino acid at 72 sites, and different ones at 68.

Hence $$68 = 140(1-e^{-k})$$

from which $$k = -\log_e\left(\frac{72}{140}\right) = 0.665$$

This result implies that, averaging over the whole α-haemoglobin chain, there are 0.665 amino acid substitutions per site between man and the carp. This figure can be converted to the number of substitutions per amino acid site per year by dividing it by the length of time during which evolution has been taking place. This is equal to twice the time since the separation of the stocks leading to these two species since evolution has presumably been acting on both lineages. It seems likely that this divergence was early in the Devonian Period, about 375 million years ago. The rate of substitution can therefore be estimated as $0.665/(2 \times 375 \times 10^6) = 8.9 \times 10^{-10}$ amino-acid sites per year.

King and Jukes[158] performed similar computations for the 9 proteins listed in Table 3.4. They compared all mammalian species for which sequence data was available, counting up the number of differences between sites, and the total number of pairs of sites which could be compared. They allowed for the possibility of the same substitution having occurred in different lineages, resulting in sequence identity when in fact two substitutions had occurred. They then estimated the proportion of substitutions per site, and divided by twice 150 000 000 years as an assessment of the length of time since the origin of mammals. The results of their computations are also shown

Table 3.4 The rates of amino-acid substitution for a variety of mammalian proteins. From King and Jukes.[158]

Protein	Total number of comparisons of amino acids	Observed number of differences	Number of differences per codon	Estimated number of substitutions per codon*	Substitutions per codon per year $\times 10^{-10}$
Insulin A & B	510	24	0·047	0·049	3·3
Cytochrome-c	1040	63	0·061	0·063	4·2
α-haemoglobin	432	58	0·137	0·149	9·9
β-haemoglobin	438	63	0·144	0·155	10·3
Ribonuclease	124	40	0·323	0·390	25·3
Immunoglobulin	102	40	0·392	0·498	33·2
Fibrinopeptide A	160	76	0·475	0·644	42·9
Bovine foetal haemoglobin	418	97	0·221	0·250	22·9
Guinea pig insulin	255	86	0·337	0·411	53·1

* Allowing for multiple substitutions, and the same substitutions in different lineages.

in Table 3.4, and it is clear that the rates of substitution differ markedly. A plausible explanation would be that selection operates at different rates in different proteins. But there are two apparently compelling reasons for dismissing this argument, and for questioning the relevance of natural selection to the evolution of protein sequences.

The first of these concerns the rates of substitutions of amino acids within homologous protein molecules in an array of species. Kimura and Ohta[154] give expressions for the estimation of the statistical errors of the rates of substitution and apply these to the α-haemoglobin chain of carp, man, horse, cow, pig, rabbit and sheep. They conclude that there is little evidence of differences between these species in their substitution rates when the statistical error inherent in the estimation process is taken into account. The rates of substitution are effectively the same in species which have been evolving separately for over 350 million years. Furthermore, the rates are the same in cytochrome-c from the silkworm, tuna, rabbit and man when they are compared with wheat.

Such constancy is difficult to reconcile with selection, for how could selection operate in such a consistent way? But if the substitution of genes were selectively neutral, the rates would be constant and depend upon the length of time involved in a much more regular way.

The second reason which can be adduced to justify considering that natural selection is irrelevant to the evolution of amino-acid sequences concerns the speed with which the substitution of novel genes must have taken place. The substitution rate for the α-haemoglobin chain between carp and man has been estimated as $8{\cdot}9 \times 10^{-10}$ per amino acid site per year. This implies that a new amino acid is substituted at a particular site every 11×10^{8} years. Kimura[150] took the average substitution rate from a series of proteins and suggested that a rough figure for a hypothetical protein of 100 amino acids in length is one substitution every 28×10^{6} years. But 20% of nucleotide substitutions, especially those at the third site, code for the same amino acid, so the nucleotide substitution rate is 1.20 times that for amino acids. Since nucleotide substitutions are the raw material of mutation, it is necessary to apply this correction factor when considering the rate of substitution of new mutant genes. Therefore, a nucleotide substitution of some kind, either synonymous or not, occurs every

$$\frac{28 \times 10^{6}}{1.2} \text{ years}$$

Muller[200] estimated that human DNA consists of 4×10^{9} nucleotide pairs, which is equivalent to $\frac{4 \times 10^{9}}{300}$ of the hypothetical 100 amino-acid proteins. If there is a substitution every $28 \times 10^6/1.2$ years when we consider one gene, taking $\frac{4 \times 10^9}{300}$ genes results in one substitution every $\frac{28 \times 10^6}{1.2} \times \frac{300}{4 \times 10^9}$ years. This equals 1.75 years.

A substitution rate of one new gene becoming fixed in the genome every two years is in marked contrast to Haldane's estimate of one every 300 generations.[120] Indeed, no mammalian species produces sufficient offspring for such a substitution rate to be possible. Kimura[150] recognized this, and suggested that such a substitution rate was possible in the absence of natural selection, provided the mutation rate was high enough. In fact, he estimated that it needed to be one mutation per genome per generation, and considered this to be not unreasonably high.

This result and the conclusion stemming from it gave rise to a storm of controversy in the pages of *Nature* and *Science*. From the flood of criticism, accusations and counter-criticism (sometimes couched in unscientific terminology—'naive panselectionist' became a term of near abuse in the years around 1970) we can select a few pertinent facts. To begin with, Kimura's result assumed that the entire genome codes for proteins which are evolving at the rate of haemoglobin, cytochrome-c, and those other structural proteins which have been sequenced. We have little idea of the validity of this assumption, but, as Maynard Smith[244] pointed out, it is difficult to see what other assumption could have been made at the time.

A much more serious criticism of the neutralist argument is that we know now that substitutions are not random within the genome, or, to be more precise, they are not random within the proteins which have so far been sequenced. Certain sites are constant between the sequences from different species, and there are more of these than would be expected from chance alone. King and Jukes[158] in a spirited attack on selectionism examined the sequence data from 25 species for which cytochrome-c had been completely sequenced. They showed that some sites were constant, others had a single substitution, some two, and so on, as shown in Table 3.5. There is clearly an excess of sites at which no substitution has taken place. They suggest that some of these are invariant for an underlying biological reason, and this is perfectly reasonable. Amino acids close to the active site of an enzyme are likely to be critical for the correct functioning of the molecule. Furthermore, it is known that when a molecule doubles back upon itself, hydrogen bonds may form between two sulphur atoms of adjacent cysteine units. A substitution at a cysteine site involved in one

Table 3.5 Number of amino acid sites at which a particular number of substitutions have taken place during the evolution of three molecules. The best available fit to the Poisson distribution is given for each molecule to show that there is always an excess of 'invariant' sites—i.e. sites which remain unchanged in their amino acid throughout the observed period of evolution. From King and Jukes.[158]

		Number of changes per site									
		0	1	2	3	4	5	6	7	8	9
Globins	observed	7	21	23	33	29	20	7	5	2	1
	Poisson	4	15	27	31	27	19	11	6	2	1
Cytochrome-c	observed	35	17	18	19	10	6	3	1	1	0
	Poisson	6	16	20	18	12	6	3	1	0·3	0·1
Specificity regions of light chains of immunoglobulins	observed	17	19	28	20	12	5	5	4	1	0
	Poisson	9	22	27	21	13	6	2	0·8	0·3	0·1

of these disulphide bonds will result in the breakage of the bond and a consequent lack of cohesion of the molecule itself. Similarly, proline is known to be an amino acid with a bend in it. It is thus particularly valuable at the 'corners' of molecules, and again a base substitution which results in a change of proline for something else, or *vice versa*, could have disastrous effects upon the conformation of the molecule.

King and Jukes pointed out that the pattern of substitutions listed in Table 3.5 somewhat resembled a Poisson distribution. A closer fit to this was obtained if certain of the invariant sites were omitted from the analysis. Since the data then fit a Poisson, or a random, distribution, they suggested that at the non-essential sites, substitution of amino acids is random. Leaving aside the valid criticism of Clarke[42] that any distribution can be successfully explained if some of the data are omitted, we can challenge the conclusion on logical grounds. King and Jukes suggest[158] that their derived random pattern 'indicates that there is very little restriction on the type of amino acid that can be accommodated at most of the variable sites'. This is not so. The fact that the substitutions fit a Poisson distribution implies that substitutions may be random with respect to location. It does not tell us anything about the *types* of amino acids which are involved.

Fitch and Markowitz[99] approached the problem of invariant sites in a more methodical way. They agreed that certain of these were constant for sound biological reasons: mutations affected the behaviour of the molecule and reduced the fitness of individuals which carried the altered genes. Others, however, were invariant simply because evolution had not got round to changing them by random processes. These sites could vary: indeed, in other species they *did* vary. But in the species which they were considering they were constant. For example, they compared the sequences of cytochrome-c from a series of mammalian species, and concluded that only about 10% of the 104 or so codons were free to vary in any one species at any one time. These 10 or 11 sites they called 'concomitantly variable codons' or covarions for short. They pointed out that the covarions in one species need not be the same as the covarions in another. When they examined the sequences of mammalian fibrinopeptide-A, they found that the number of covarions appeared to be about 18 out of 19 codons.

Fitch and Markowitz then proceeded to analyse the number of amino acid differences in the cytochrome-c and fibrinopeptide-A from the horse and the pig.

They found there to have been five amino-acid substitutions in the cytochrome-c molecule, and 13 in the fibrinopeptide-A. This gives substitution rates of 0.048 and 0.684 fixation per codon in the two lines since their evolution from a common ancestor. Such results suggest that the evolutionary rate has been very different in the two genes, which is not to be expected upon a random mutation and substitution model. However, when the number of substitutions per covarion is calculated, the picture changes dramatically. There are now 5 amino-acid substitutions among 10 covarions in the cytochrome-c genes, and 13 out of 18 in the fibrinopeptide-A. These rates are not statistically different, and Fitch and Markowitz offer this as evidence that amino-acid, and hence nucleotide, substitution is random within covarions. The result is appealing. A certain number of sites are of major importance to the correct structure and functioning of the molecule, but the remaining sites are less critical. Within these sites, it does not matter much which amino acid is present, and so the nucleotides are free to vary at random.

King[156] points out the inadequacy of this kind of argument. Allowing for invariant sites in the same way, he estimates the number of substitutions to the cytochrome-c sequence of wheat, fruit fly and man as follows:

Wheat to man	46
Wheat to fruit fly	63
Man to fruit fly	29

He agrees that the estimates are likely to be inaccurate since they assume all covarions are equally variable. However, using these figures, we can produce an evolutionary tree, as in Fig. 3.5, which implies that fruit flies have been evolving four times as rapidly as man. This seems unlikely under a neutralist view such as that propounded by Jukes[139] in the same volume: 'The evolutionary clock ticks slowly in proteins, independent of speciation, generation time or gene duplication.'

Wheat – Man = 46 substitutions

Wheat – Fruit fly = 63 substitutions

Man – Fruit fly = 29 substitutions

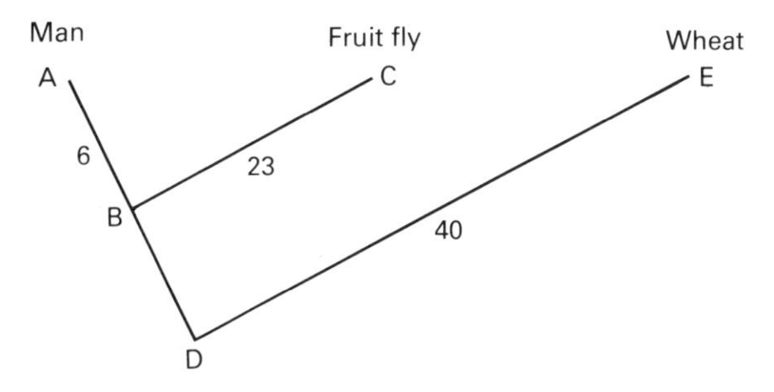

Fig. 3.5 Evolutionary tree of wheat, man and fruit fly based upon the rates of substitution in covarions, derived as in the hypothetical example of Fig. 3.1.

The difficulty has arisen because there seems to be no way of testing whether a substitution rate of 6 amino acids per covarion in the lineage leading to man is statistically different to one of 23 in the fruit fly line. This problem had initially arisen during the analysis of King and Jukes.[158] They compared the number of substitutions per codon per year in a variety of proteins and concluded that they were sufficiently similar for an average rate to be estimated. Richmond[226] disputed this, and suggested that in his opinion they varied rather a lot. It now seems clear that Richmond was correct, for Kimura[151] gave a formula for the estimation of the variance of a substitution rate. Applying this to a series of evolutionary lineages (Table 3.6) shows there to be significant differences between them. The rates of substitution are thus not uniform when we consider the whole molecule.

However, this argument relates to the basic substitution rate. When

Table 3.6 The rates of substitution and the errors of their estimation during the evolution of various haemoglobins. The length of time since the separation of the species under comparison is shown in millions of years in column T. From Kimura.[151]

Comparing	With	Rate of substitution	T
Carp α	human, mouse, rabbit, bovine α	8·9±0·5	7·5
Human α	horse, bovine, pig, sheep α	8·8±0·9	1·6
Mouse α	human, horse, bovine, pig, rabbit, sheep α	10·9±0·9	1·6
Human β	horse, pig, sheep, bovine β	11·9±1·0	1·6
Mouse β	human, rabbit, horse, pig, bovine β	14·0±1·2	1·6
Human β	human, mouse, rabbit, horse, bovine, carp α	8·9±0·4	9·0
Rabbit β	human, mouse, rabbit, horse, bovine, carp α	9·2±0·4	9·0
Human β	lamprey globin	12·8±1·4	10·0

one compares the rate of substitution per covarion, the variance is not so easily estimated. There is a statistical error inherent in the estimation of the number of covarions, which in turn will reduce the accuracy of the computed substitution rate. Until the problems are solved, the arguments for and against the neutralist view will continue.

However, King's[156] result suggests that the rates of substitution per covarion *may* differ within the same protein in different species. Such a result would be predicted under a selectionist model, although we have seen that the rates of substitution are supposed to place an intolerable burden upon the reproductive systems of the species concerned. Perhaps it is the model that bridges the gap between rates of substitution and the reproductive strategy of the species which is wrong.

Maynard Smith[224] and O'Donald[206] both attempted to provide a more rational model for the observed high rate of substitution. Both offer a selectionist view, but are equally unrealistic over-simplifications.

Further problems arose when Clarke[43] showed that there are constraints upon the *types* of substitution that have taken place during evolution. He counted the number of substitutions that had occurred in a series of proteins from Dayhoff's 1968 Atlas. He then standardized this figure to allow for the relative frequencies of amino acids, and also the differences in the probability of mutation from one to another. This gave him a series of connected numbers of substitutions between pairs of amino acids that had taken place during the evolution of those proteins that he had examined.

Sneath had earlier[245] analysed 134 chemical characteristics of the amino acids involved in protein structure. Using a technique from numerical taxonomy, he derived coefficients of resemblance between all possible pairs of these amino acids. Clarke produced a graph to show the relationship between the two sets of data, including only those pairs of amino acids that differed by one-step mutations in the trinucleotides. His graph is shown in Fig. 3.6, and it is at once apparent that the more frequent substitutions involve amino acids that are chemically more similar. Such a result would not be expected if substitutions were random. The chemical properties of the amino acid should not affect its likelihood of being involved in a substitution.

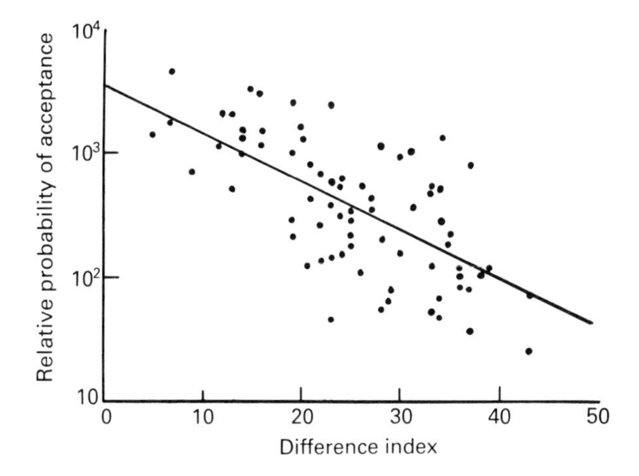

Fig. 3.6 The relationship between the probability that one amino acid will be substituted for another in a polypeptide, and the similarity between those amino acids. After Clarke.[43]

Such a result is entirely consonant with natural selection: the more different amino acids are less likely to be successfully substituted, for their chemical properties may interfere with the correct functioning of the molecule. This interference will in turn give rise to a reduction in the fitness of the individuals concerned.

This further invalidates the analysis of King and Jukes.[158] Substitutions are not random with respect to site. Neither are they random with respect to the amino acids which are involved. This exercise gives us an insight into the way in which a scientific argument may develop. An initial observation, that of the apparently good fit to a Poisson distribution, gave rise to ideas. These were dissected and discussed at length (and with acrimony, for scientists are human like everyone else). Eventually some semblance of order began to emerge, and the heat went out of the situation. Nevertheless, unsolved problems remain, although it seems to be necessary to search elsewhere for conclusive evidence pertaining to the relative importance of selection and random events in the evolutionary process. This we shall do in the ensuing chapters.

4

Genetic Polymorphism

THE AMOUNT OF POLYMORPHISM

In the previous chapter, we discussed how recent developments in molecular genetics have begun to make real contributions to our understanding of evolution at the level of the individual gene. Recent technical developments in biochemistry have also produced important advances in evolutionary genetics, particularly in the estimation of the amount of heterozygosity in the average individual and the proportion of loci that are polymorphic in the average population. The results of this work have lead to a reappraisal of genetic polymorphism.

The question of the amount of variation in natural populations has taxed geneticists for many years. The problem was traditionally approached by enquiring what might be the number of segregating loci in an average population. Earlier estimates varied widely, ranging from a very small proportion[60] to about 50%.[266] These estimates were based upon rather complicated arguments concerning the number of deleterious, or fitness-reducing, mutant genes present within a single individual. Such indirect approaches were inevitable, because, while it was difficult enough to estimate the number of segregating loci in a population, it was virtually impossible to assess how many were not segregating. The biochemical technique that has allowed a more direct approach to the problem is electrophoresis, and it is worth devoting a few lines to this.

The principles of electrophoresis are similar to those of chromatography, and they are fairly simple. A gel matrix is produced, usually from starch, polymerized acrylamide, agar or agarose, in some appropriate buffer. In population genetics, the first two types of gel are most frequently used. They are normally made in thin rectangular slabs, although polyacrylamide gels can be made in small, pencil-shaped cylinders. The material to be assayed is homogenized, and either placed on the end of a polyacrylamide gel, or absorbed on a square of filter paper and stuck into the middle of a starch gel. A voltage differential is then applied across the gel and, under its influence, the molecules present in the homogenate may move. The distance that they move within a given span of time depends upon several factors, including the temperature and pH of the gel, the concentration of the gel matrix itself and, of more importance to geneticists, the size, shape and electric charge of the molecules. Identical molecules should move at the same rate, provided that the gel is uniform in consistency.

After a time, the current is stopped, and the gel is stained. This may be direct staining, such as with Coomassie Blue dye for protein or Cresol Red dye for lipids. Alternatively, an indirect colour-producing reaction may be contrived, a method that is particularly useful for detecting enzymes at low concentrations. Using either technique, a series of bands are caused to appear in the gel at the position to which the molecules have moved (see Fig. 4.1). Some of the bands are present in the same position for all individuals of a species; others are variable. An example of a variable enzyme is shown in Fig. 4.1a, which depicts a gel that has been 'stained' for esterase enzymes, using homogenized larvae of the kelp fly (*Coelopa frigida*). Genetic analysis by breeding shows that the ability to produce each band on this gel is vested in a different allele at the same esterase locus. Two bands are produced in a heterozygote because two alleles are present in the fly, and so both enzymes are represented in the gel. In normal terminology the enzyme alleles are recorded by their mobility relative to one standard allele. In a simple two allele situation, the enzyme moving further in a given time is called the 'fast' enzyme F, because it moves faster through the gel.

Some enzymes are dimeric in structure, with the alleles producing sub-units rather than the entire molecule. Such situations can result in a much more complex gel pattern (zymogram). Fig. 4.1b shows a zymogram of alcohol dehydrogenase from *Drosophila melanogaster*. There are two alleles, and heterozygotes produce two different sub-units. These combine with sub-units which may or may not be like themselves. Consequently, three enzymes are produced, one of them being a hybrid, and three bands appear upon the gel. A further complexity is introduced because the chemical nicotinamide adenine dinucleotide (NAD) is present in the homogenate. This acts as a cofactor, and also binds to the alcohol dehydrogenase (Adh) molecule changing its mobility. Up to eight or nine bands can be produced this way.

The genotype of an individual can be determined fairly easily for a protein using electrophoresis, and this allows estimates of heterozygosity to be made quite objectively. If a single band is present in all individuals of a population, it is likely that this reflects a locus that is homozygous in all individuals, although it is always possible that alleles are present that code for proteins having similar mobilities. Having regard for this, however, the number of invariant and variable loci can be counted and the proportion of polymorphic loci can be obtained.

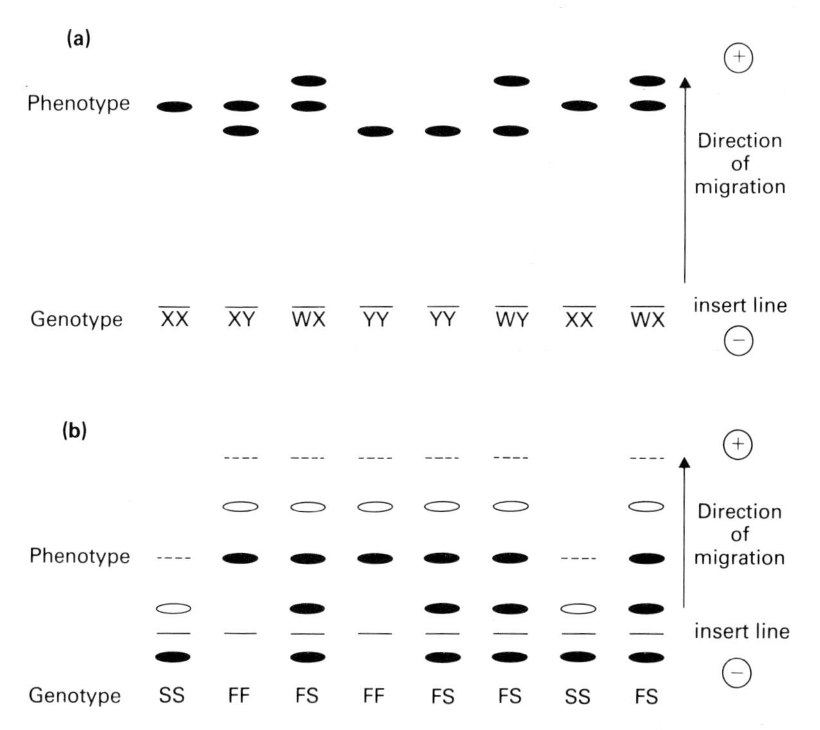

Fig. 4.1 Zymograms of (**a**) larval esterases of *Coelopa frigida* and (**b**) alcohol dehydrogenases of *Drosophila melanogaster.* Note the two-banded nature of heterozygotes in the former. The complex patterns in the latter result from firstly the presence of a hybrid dimeric enzyme form in the heterozygote, and the presence of satellite bands caused by the binding of additional biochemical molecules to the enzyme, with a consequent alteration in mobility.

Pioneer studies of this kind were made by Lewontin and Hubby[126, 176] and by Harris.[124] These authors used rather complex methods so we will describe the techniques used by later workers and then return to the deductions made by Lewontin and Hubby which are supported by subsequent research.

A typical study is that of Selander and Yang[235] who report upon the electrophoresis of several hundred house mice (*Mus musculus*) collected in ten large chicken barns on a farm in California. Large numbers of individuals were assayed for a variety of enzymes and other proteins, which are listed in Table 4.1. The inheritance of many

Table 4.1 Electrophoretic variation detected in a survey of 35 proteins from a Californian population of the House Mouse (*Mus musculus*). The variation reported can be explained by 40 loci, of which 12 (or 30%) are polymorphic. From Selander and Yang.[235]

Protein	Number of loci	Phenotypic variation
Enzymes		
Alkaline phosphatase	1	polymorphic
Esterase	4	polymorphic
Plasma esterase	3	monomorphic
Erythrocyte esterase	1	monomorphic
Aldolase	2	monomorphic
Alcohol dehydrogenase	1	monomorphic
Lactate dehydrogenase	2	monomorphic
Lactate dehydrogenase regulator	1	polymorphic
Malate dehydrogenase	3	monomorphic
Xanthine dehydrogenase	1	monomorphic
Hexose-6-phosphate dehydrogenase	1	polymorphic
Glucose-6-phosphate dehydrogenase	1	monomorphic
6-phosphogluconate dehydrogenase	1	polymorphic
Isocitrate dehydrogenase	2	1 polymorphic 1 monomorphic
Indophenol oxidase	1	monomorphic
Phosphoglucomutase	2	1 polymorphic 1 monomorphic
Phosphoglucose isomerase	1	monomorphic
Fumarase	2	monomorphic
Non enzymatic proteins		
Haemoglobin (β chain)	1	polymorphic
(α chain)	1	monomorphic
Non-haemoglobin erythrocyte protein	1	monomorphic
Pre-albumin	2	1 polymorphic 1 monomorphic
Albumin	1	monomorphic
Transferrin	1	monomorphic
Plasma protein	3	monomorphic

of the enzymes that occur in mice had already been determined, so they were able to record individuals as homozygous or heterozygous by reference to the earlier work. Loci were recorded as homozygous if every individual possessed the same band, and such loci were assumed to be monomorphic. When segregation was apparent on the gels, the protein locus concerned was recorded as polymorphic. The results of the assays are given in Table 4.1. Selander and Yang found that 12 out of the 40 loci were polymorphic in the population of mice living in the 10 barns, so approximately 30% of the loci were polymorphic. Furthermore, Selander and Yang were able to show that the average individual was heterozygous at 11% of these loci. These figures are in close agreement with the earlier estimates of Lewontin and Hubby[126] and Harris.[124]

A wide range of species have now been examined for polymorphic proteins, and some of the results are shown in Table 4.2. Species that have arisen relatively recently, such as some of the *Drosophila*, have amounts of heterozygosity similar to species that can best be described as phylogenetic relics. King crabs, for example, have remained virtually unchanged since the Devonian period, and yet have apparently continued to maintain high levels of heterozygosity. This must mean that balancing forces such as mutation or selection have been acting to maintain the variation in these species, for otherwise

Table 4.2 Estimates of the amount of genetic polymorphism in natural populations of a series of widely separate species.

Species	Number of populations	Number of loci	Proportion of loci polymorphic per population	Reference
Homo sapiens	1	71	0·28	Harris and Hopkinson (1972)
Mus musculus	4	41	0·29	Selander *et al.* (1969)
Peromyscus polionotus	7	32	0·23	Selander *et al.* (1971)
Drosophila pseudoobscura	10	24	0·43	Prakash *et al.* (1969) Prakash *et al.* (1973)
Drosophila melanogaster	1	19	0·42	Kojima *et al.* (1970)
Limulus polyphemus	4	25	0·25	Selander *et al.* (1970)

many of the loci would have drifted or been selected to homozygosity during the aeons of time between their origin and the present day.

The significance of the results in Table 4.2 is considerable. There is clearly a large amount of genetic polymorphism in many populations, and in a wide range of species. Some of this polymorphism may be transient. The alleles that are segregating could be in the process of increasing or decreasing in frequency either selectively or neutrally. However, there is another class of polymorphisms which seem to be very long-lasting, or even permanent. A clear example of a very long-lasting polymorphism comes, not from electrophoresis, but from a study of shell colours and banding patterns in the land snail *Cepaea nemoralis*. Ancient shells have been collected from the ruins of villages dating back to Neolithic Britain.[63] Several different banding phenotypes have been found among this material, and present day populations at, or near to, the same sites are still polymorphic for the same phenotypes. This suggests that the polymorphism may have persisted in these populations for at least 6500 years.

We can infer that most of the visible polymorphisms in *C. nemoralis* and *C. hortensis* are even older than this. The shells of the two species vary in ground colour. There is also variation in the colour, number, and position of the bands, and the colour of the peristome. In every case which has been analysed, the inheritance of these characters is identical in the two species, as are the dominance relationships of the alleles involved. It seems that these polymorphisms are more ancient than the separation of the species themselves.

There are enzyme polymorphisms which may be equally ancient. Hubby and Throckmorton[127] list the alleles at the xanthine dehydrogenase locus in a series of species of *Drosophila*. Several of these alleles appear to be shared by the species group which includes *D. viridis*, *D. americana* and *D. texana*. Members of the related phylad which includes *D. littoralis* and *D. borealis*, as well as other species, possess some of these alleles in addition to several particular to their own group. The simplest hypothesis might be that these alleles are older than the evolutionary processes that produced the *Drosophila* species involved. How these polymorphisms have remained in the face of selection and genetic drift is clearly a matter of great theoretical and practical interest.

THE MAINTENANCE OF POLYMORPHISMS

Several mechanisms have been proposed which will operate to actively maintain a polymorphism in a natural population. Most of them have been explored theoretically, and some have been subjected to more or less rigorous experimental research. The two simplest are heterozygous advantage and frequency-dependent selection. In the former, the heterozygotes possess some advantage over the two homozygous genotypes. This advantage may be manifest in one of several ways: superior viability, fertility or fecundity of the genotype itself, or a longer reproductive life, for example. As a result, the heterozygotes show a superior fitness, and leave relatively more offspring than other genotypes. Consequently, both alleles are passed on to the next generation, and the polymorphism is maintained. Frequency-dependent selection, on the other hand, involves situations where the fitness of a genotype (sometimes of a gene) increases as its frequency declines. Consequently, the genes composing that form are produced in relative excess when they become rare in the population.

There are other, less well-studied mechanisms. Density-dependent selection is the situation where the fitness of a genotype is related to its density; this will also maintain a polymorphism under certain conditions. So will differential selection on the various genotypes at different stages of the life cycle, and differential selection between the sexes can act to maintain segregation as well.

HETEROZYGOUS ADVANTAGE

Heterozygous advantage is probably the most widely quoted method of maintaining a polymorphism, and simultaneously the least demonstrated phenomenon in population genetics. When a heterozygote is superior in fitness to both homozygotes, then a disproportionately large proportion of heterozygous offspring survive to reproduce and inevitably both alleles are maintained in the population. This can be demonstrated using the following simple model.

Let there be two alleles A_1 and A_2 segregating in a large, randomly mating population, and let the fitnesses of the three genotypes be as shown in Table 4.3. The parameters s and t lie between zero and unity; they need not be equal in value. After a generation of random mating, and applying the appropriate fitnesses, the frequency of the A_2 allele changes from q to:

$$q' = \frac{q - tq^2}{1 - sp^2 - tq^2}$$

The change in gene-frequency is given by:

$$\begin{aligned} \Delta q &= q' - q \\ &= \frac{q(1-q)(s - sq - tq)}{1 - sp^2 - tq^2} \end{aligned}$$

Table 4.3 A simple predictive model for heterozygous advantages. For further details, see text.

Two autosomal alleles segregate in a large, random mating population:

	frequency of $A_1 = p$	frequency of $A_2 = q$	$p+q=1$
Possible genotypes	A_1A_1	A_1A_2	A_2A_2
Frequency after zygote formation	p^2	$2pq$	q^2
Relative fitnesses	$1-s$	1	$1-t$
Genotype frequencies after selection	$p^2(1-s)$	$2pq$	$q^2(1-t)$

Total of individuals after selection $= 1-sp^2-tq^2$

Frequency of A_2 allele after selection $= q' = \dfrac{pq+q^2(1-t)}{1-sp^2-tq^2} = \dfrac{q-tq^2}{1-sp^2-tq^2}$

$$\text{Change in gene-frequency} = \Delta q = q'-q = \frac{q-tq^2}{1-sp^2-tq^2}-q$$

$$= \frac{sq-tq^2-2sq^2+tq^3+sq^3}{1-sp^2-tq^2}$$

$$= \frac{q(1-q)\,(s-sq-tq)}{1-sp^2-tq^2}$$

$\Delta q = 0$ at equilibrium

$\Delta q = 0$ if $q = 0$, $1-q = 0$ or $s-sq-tq = 0$

i.e. when $q = 0$, $q = 1$ or $q = \dfrac{s}{s+t}$

At equilibrium, gene-frequencies do not change from one generation to the next, so $\Delta q = 0$, and therefore

$$q = 0$$
$$1-q = 0$$
$$\text{or} \quad s-sq-tq = 0$$

The first two of these represent trivial equilibria where the population is homozygous for A_1 and A_2 respectively. The third, however, gives an equilibrium frequency of:

$$\hat{q} = \frac{s}{s+t}$$

This is an interesting result for it implies that there is a non-trivial equilibrium point which depends solely upon the relative sizes of s and t. For example, if we put $s = 0.1$ and $t = 0.3$, the fitness of the homozygotes are 90% and 70% of the heterozygotes. These selection coefficients should give rise to an equilibrium at $\hat{q} = 0.25$.

It is instructive to plot the relationship between the change in gene-frequency under heterozygous advantage and the gene-frequency itself in the parental population. Fig. 4.2 shows that graph using the values of s and t from the previous paragraph. As would be expected,

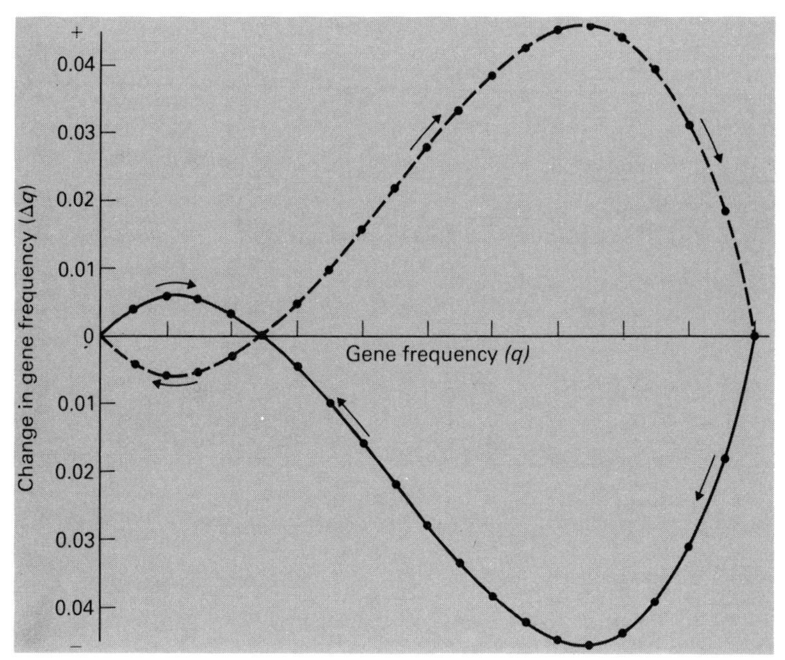

Fig. 4.2 The relationship between the initial frequency of a gene q and the change in its frequency Δq under simple models of heterozygous advantage (•———•) and heterozygous disadvantage (•- - - - -•). The direction of gene frequency change is indicated by the arrows.

the line relating the two parameters passes through $\Delta q = 0$ at $q = 0$, $q = 1.0$ and $q = 0.25$. The significance of the graph is that Δq is positive when q is less than 0.25, and negative when it is greater. This implies that there is an increase in the value of q when it drops below the equilibrium point, and a decrease when it rises above. The gene-frequencies thus converge upon the equilibrium point which is stable, for perturbations are followed by restitution to the equilibrium situation. Of course, restitution of gene-frequency cannot take place if the perturbations are sufficiently extreme to eliminate one or other allele—these are the situations represented by the trivial equilibria.

There is a direct analogy between this genetic model and physical equilibria as typified by a symmetrical cone (see Fig. 4.3). When the cone lies upon its side it is in neutral equilibrium, for slightly disturbing the cone results in its moving to a new position where it will remain until the next disturbance. This corresponds to the genetic equilibrium that occurs when s and t are both equal to zero,

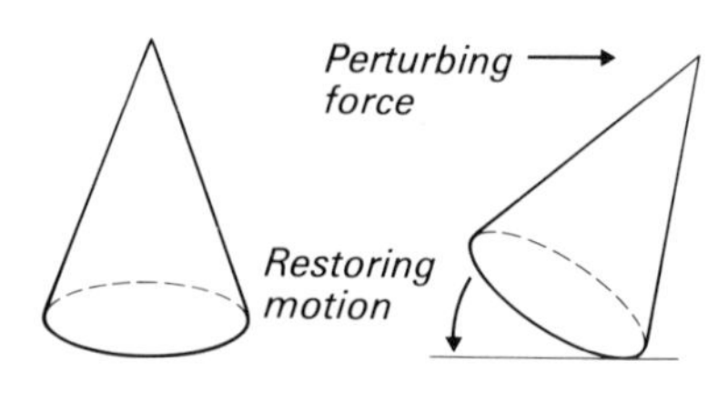

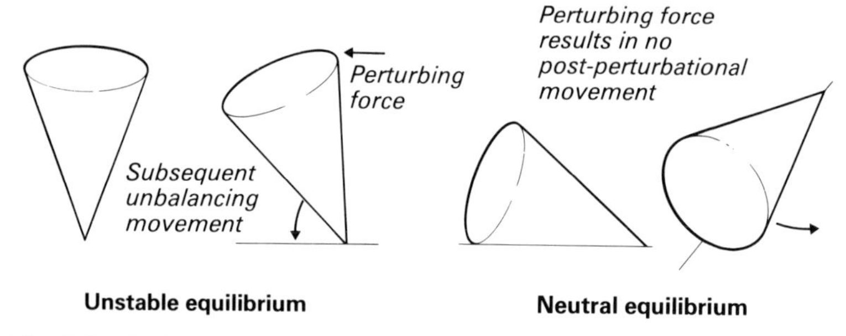

Fig. 4.3 Stable, unstable and neutral equilibria as depicted by a cone placed upon a flat table.

and is the same as the billiard ball analogy to the Hardy-Weinberg law in Chapter 2. The second position of the cone is a stable state equilibrium, for a perturbation of the cone here will be followed by restitution, unless it is tilted too far when it will overbalance with no possibility of restitution. The third is an unstable equilibrium, for the slightest disturbance will be followed by immediate overbalancing, again with no chance of restitution.

Our model of heterozygous advantage is essentially the same. When $s = 0$ and $t = 0$, the system is neutral and the gene-frequencies will drift up and down until eventually one or other allele is fixed. If s and t lie between 0 and 1 the equilibrium is stable with perturbation leading to restitution. If s and t lie between 0 and -1, the heterozygote has an inferior fitness to the two homozygotes. There is still an equilibrium at $\hat{q} = s/s+t$, but it is unstable. The graph relating Δq to q is shown by the broken line in Fig. 4.2. When $q < \frac{s}{s+t}$ then $\Delta q < 0$, and conversely when $q > \hat{q}$ so $\Delta q > 0$. Thus, a slight perturbation of gene-frequency is followed by a progressive movement towards fixation of whichever allele is present in above-equilibrium frequency.

There are very few examples of heterozygous advantage that have been fully researched. True heterozygous advantage relates to alleles at a single locus, and a great deal of study of the genotypes and the mechanistic connection between the gene products and the selective agents is needed to prove it. As yet, few genetic systems have been studied in sufficient depth to allow such an analysis to be made. The most widely quoted, and easily the best understood, concerns the disease called sickle-cell anaemia in man. This is now known to be caused by the substitution of valine for glutamic acid at site number six of the β-haemoglobin polypeptide. Individuals who are homozygous for this altered gene have sickle-shaped red blood corpuscles and very rarely survive to reproductive age. The heterozygote is usually normal (so the sickle-cell gene is recessive), but when the oxygen concentration in a tissue declines, the erythrocytes collapse into their characteristic sickle shape. Under normal conditions, however, heterozygous people are reasonably healthy.

The gene causing this disease is present at a high frequency among many of the peoples of Africa and southern Asia (see Fig. 4.4a), despite its apparent selective disadvantage due to the increased morbidity of the homozygotes. For many years, the reasons for this remained shrouded in mystery, until Allison completed a series of brilliant studies which proved that this disadvantage was balanced by the effects of subtertian malaria (reviewed by Allison[4]). Subtertian malaria is caused by the sporozoan parasite *Plasmodium falciparum*, and its distribution around the world is shown in Fig. 4.4b. Allison produced two crucial pieces of evidence. Firstly, he examined the relationship between the incidence of malaria and the sickle-cell genotype in a sample of children under six years old from Uganda. He found[2] that of 43 children who were heterozygous for the sickle-cell gene, only 12 had *P. falciparum* in their blood. On the other hand, 113 out of a sample of 247 children with normal haemoglobin contained the parasite. The difference between these figures is statistically significant. Furthermore, only 4 out of the 43 heterozygotes showed heavy infections compared with 70 out of the 247 normal children. Again this is a statistically significant difference.[2]

These results indicate not only that heterozygotes are less likely to be infected with the malarial parasite, but also that any infection is likely to be less severe. These results have been confirmed by a series of workers (listed by Levitan and Montague[172]), and there is no doubt that Allison has found the balancing force. Individuals who are homozygous for the sickle-cell gene have a reduced viability due to

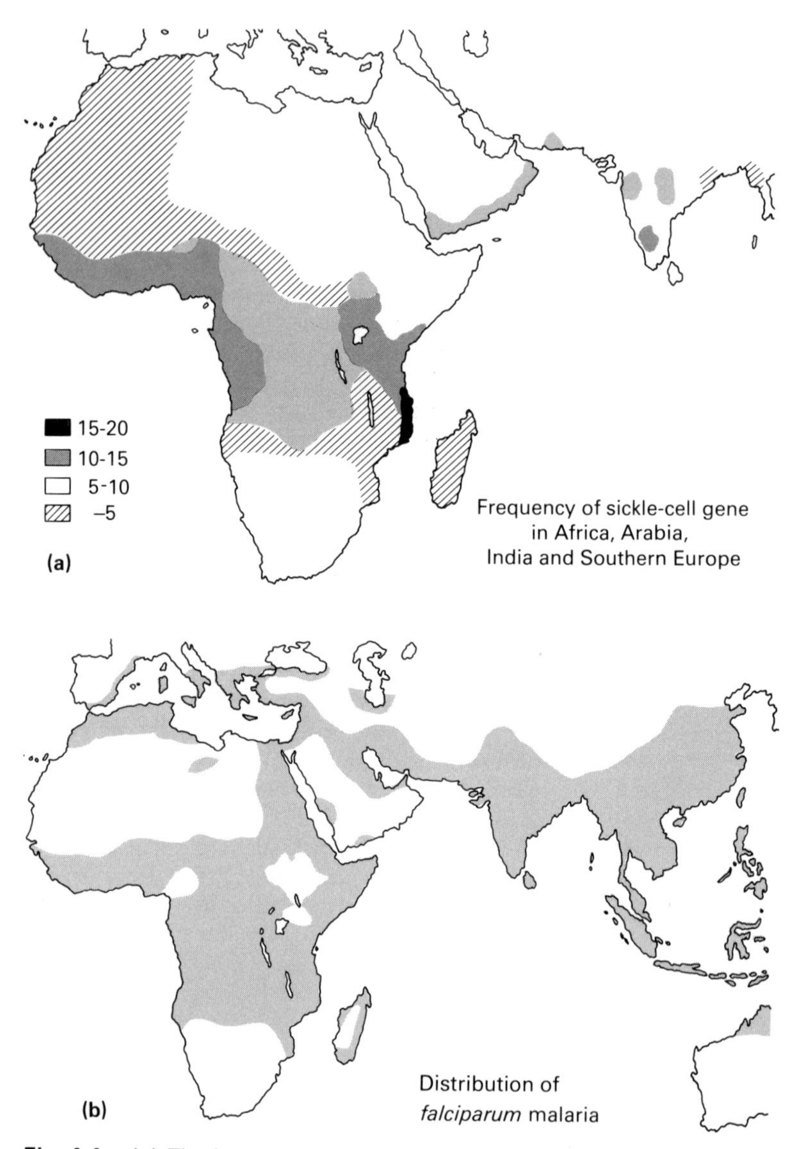

Fig. 4.4 (**a**) The frequency of the gene for sickle-cell anaemia in Africa and south western Asia. (**b**) The distribution of *falciparum* malaria in the same regions. Note the considerable areas of agreement in the distributional patterns, especially in central Africa. (From Allison, A. C. (1961). In *Genetical Variation in Human Populations.* Pergamon, Oxford.)

the anaemia and its secondary effects. Normal individuals, on the other hand, are more prone to malaria, and have an enhanced mortality as a result. The heterozygotes are at an advantage over both, and so the polymorphism is maintained. As might be expected, when a population leaves the malarial area, there is a decrease in the frequency of the sickle-cell genes in ensuing generations,[3] because the situation is now one of directional selection against a homozygote.

It appears that the solubility of deoxygenated sickle-cell haemoglobin is much less than normal haemoglobin.[210] In the oxygenated state they are equally soluble, but it seems that when the oxygen concentration declines the sickle-cell haemoglobin crystallizes out of solution and distorts the red cell into its characteristic shape. This is followed by a deterioration in the oxygen transport system, which is further aggravated by increased rates of phagocytosis of the deformed erythrocytes. Miller *et al.*[188] suggested that red cells containing *P. falciparum* adhere to the walls of small blood vessels while the parasites mature. In these situations the oxygen tension is low and sickle-cell haemoglobin begins to crystallize, the erythrocytes then sickle and are phagocytosed. The parasite is thus destroyed before it reaches maturity, and so cannot reproduce to infect more erythrocytes. As a result, the parasite is less likely to become established in the bloodstream of a heterozygote. Should a heterozygote become infected, however, it is unlikely that the parasite will attain a sufficient level to cause a severe case of malaria.

There are other examples of heterozygous advantage which stem from the interaction between malaria and genetic polymorphisms. There is another mutant form of the β-haemoglobin gene called haemoglobin c, which is also resistant to malaria when present in the heterozygous form. The circumstantial evidence for the glucose-6-phosphate dehydrogenase polymorphism is good, and Ceppellini[35] relates β-thalassaemia to malaria as well.

Evidence is also accumulating from a variety of sources that the gene conferring resistance to Warfarin upon rats might be at a selective advantage when it is present in the heterozygous conditions (briefly reviewed by Greaves[114]). Rats that are homozygous for the resistance genes are less efficient at reducing vitamin K epoxide back to the vitamin. Consequently, they need external sources of vitamin K to keep them healthy. In areas where Warfarin is laid down, non-resistant rats are liable to die from the chemical's inhibition of the blood clotting mechanism. However, this whole field warrants further study at a truly ecological level for it promises to give a further example of heterozygous advantage.

THE RELEVANCE OF HETEROZYGOUS ADVANTAGE TO PROTEIN POLYMORPHISM

We are now in a position to consider the possibility that the enzyme polymorphisms which have been studied by Hubby and Lewontin,[126, 176] Harris,[124] Selander and Yang,[235] and so many others, are maintained by heterozygous advantage. It is possible to estimate the 'average' fitness of a population ($\bar{w}$) by adding together the products of the genotype frequencies and their relative fitnesses. In the model outlined in Table 4.3, this is given by:

$$\begin{aligned}\bar{w} &= p^2(1-s)+2pq+q^2(1-t)\\ &=1-sp^2-tq^2\end{aligned}$$

We have little or no idea of the fitnesses likely to be associated with the various genotypes at a polymorphic enzymes locus. But, if we assume that the homozygotes are at a disadvantage of 2% compared with the heterozygote, we can put $s = t = 0.02$. This value of selection would be too small to be detectable in any but the most intensive of studies, and yet is sufficiently large to overcome random forces in all except the smallest populations. Under such a selection regime, an equilibrium will be attained where $p = q = 0.5$. Substituting the values of the selection coefficients, and the equilibrium gene-frequencies into the 'average fitness' equation gives $\bar{w} = 0.99$. If we think in terms of selection acting upon the immature stages of the organism before reproductive age is attained, this result means that 1% of the population of juveniles will die as a result of selection at the locus in question before reaching maturity.

Lewontin and Hubby[176] pointed out that there are probably about 6000 loci in the genome of *D. pseudoobscura*. Their results suggested about one third of all loci to be polymorphic. They had no evidence at all pertaining to the average fitnesses of the genotype at these segregating loci. Suppose, however, that all operate under the system of selection involving a 2% disadvantage to the homozygotes, then, for every locus, the population mean fitness would be reduced to 99%. If selection acts separately and independently at every locus, the mean fitness would be reduced to 0.99^{2000} by the 2000 polymorphic loci. The population fitness is thus about 2×10^{-9} which means that only two *D. pseudoobscura* survive out of every thousand million zygotes that are produced. This figure is unbelievably low, and there can be few if any species that can reproduce at such a rate.

What, then, has gone wrong? Lewontin and Hubby point out that *D. pseudoobscura* is a species that occurs in very substantial numbers, so that the population sizes are too large to be affected by random genetic drift. It seemed unlikely to them that mutation and migration would be responsible for such a high level of polymorphism. But the traditional multiplicative models of heterozygous advantage could only maintain so much polymorphism if reproductive rates were tens of thousands of times higher than reasonable ecological studies suggested them to be. How can we escape from the dilemma? The starting point must be a re-assessment of the results of the electrophoresis, and then a re-appraisal of the deductions drawn therefrom.

We will begin by looking again at the estimate of 30%, and the way in which it was derived. Firstly, the proteins that have been assayed may not be a random sample of the genome. Secondly, the estimate of the number of polymorphic loci may itself be biased or inaccurate. Thirdly, the polymorphisms may not be selectively based, but depend upon mutation and random genetic drift. Fourthly, selection may be acting but the model based upon heterozygous advantage could be wrong.

The first two of these will affect the quantification of heterozygosity, whereas the latter two affect the conclusions stemming from the estimates of heterozygosity. We will now consider them all in turn.

The randomness of the loci surveyed

Let us consider whether the proteins that have been assayed are a random sample of loci in the genome. It is possible that some classes of protein are more polymorphic than others. Kojima, Gillespie and Tobari[161] first drew attention to this. During a survey of several species of *Drosophila* from around the world, they divided their enzyme loci into two classes; those which were involved in the metabolism of glucose, and those which were not. They found a lower level of polymorphism among the former. Selander and Yang[235] did not find this in the house mouse, nor did Nair *et al.*[203] with *Drosophila mesophragmatica*.

In a major survey of published records, Johnson[133] compared enzymes which utilized internally regulated substances with others utilizing substrates that are more likely to stem from the external environment, such as amylases or esterases. He estimated the number of alleles at a locus with a frequency greater than 1%, in samples composed of at least 100 individuals, from as many studies as he could find. The 'internal' enzymes averaged fewer alleles per locus than the 'externals'. He also showed[133] that every one of 24 samples of *Drosophila* from a variety of species originating in widely different regions of the world, showed a higher mean number of alleles per locus for its 'external' enzymes than for its 'internals'.

Subsequently Gillespie and Langley[109] revised the data of Kojima *et al.*[161] by separating the enzymes into those that can only utilize a single substrate and those that can act upon several. They showed from a large amount of data, including that of Selander and Yang,[235] that the 'restricted' enzymes were less polymorphic than the 'less restricted'. On the whole, the single-substrate enzymes tend to operate upon substances produced within the organism, whereas the more catholic enzymes are involved in the metabolism of externally derived materials. Although this is not a general rule, there is a fairly close relationship, and Gillespie and Langley[109] seem to have resolved some of the anomalies that the earlier analysis had disclosed.

The differences in the amounts of polymorphism can be explained by considering the probable function of the enzymes themselves. Any enzyme associated with a biochemical pathway that involves internally-produced substances is likely to be presented with more predictable quantities of its substrate than an enzyme concerned with the metabolism of material from the environment. The latter will be presented with a more varied substrate, both in terms of quantity and quality. Consequently, a more flexible approach to the enzymic reactions may be advantageous when the substrates are externally derived. Similarly, if only a single substrate is provided for the enzyme, there may be a single 'best' genetic form of the enzyme for coping with it. A multiplicity of substances may favour a more flexible response—for example, if a different allelic enzyme were most suited to each of the substrates, then a polymorphic situation might be the best for that species.

Whether or not this is the correct explanation for the results provided by Johnson[133] and Gillespie and Langley,[109] an additional point must be made. Consistent differences in the amount of polymorphism between discernibly different classes of enzyme must be selectively based. It is almost impossible to account for such consistencies on a basis of random effects.

However, while these results argue strongly in favour of a selective basis for some of the enzyme polymorphisms, they leave us even more confused. It is clear that enzymic loci give an estimate of 30% heterozygosity. But this estimate itself is based upon heterogeneous enzyme types. It is not presently possible even to begin to compute a more reliable frequency of polymorphism until more is known concerning the proportion of total enzyme loci that are associated with internal or external, single or multiple substrates. Furthermore, enzymes are not the only proteins to be coded by chromosomal DNA. There are a variety of 'structural' proteins such as collagen, keratin and crystallins, that have been studied rather little by population genetics. Day[68] and Day and Clayton[69] suggested there to be very little intra-specific variation among these proteins that could be ascribed to genetic polymorphism. If this is so, then there are undoubtedly many fewer polymorphic loci among such proteins, but again the relative proportion of the genome that codes for enzymic or 'structural' proteins remains a mystery.

The assumption that the loci studied by Hubby and Lewontin, Harris and so many others are a representative sample of the genome as a whole is thus a matter for conjecture. No doubt the next few years will cast further light upon the problem, but at present we can allow neither for the relative abundance of the different classes of loci nor for the variation in polymorphism between them. We can do no more than assume that the estimates approximate fairly closely to the average situation, and bear in mind that this is no more than an assumption.

The accuracy of the estimate of heterozygosity

Let us turn now to the second criticism of the electrophoresis-based methods of estimating polymorphism. It is quite possible that the results may be inaccurate even within the proteins that are being studied. If two different alleles produce a band at the same position in a gel, they are regarded as one and the same. A recent study by Bernstein *et al.*[13] has shown that such iso-alleles exist. They examined a series of alleles at the xanthine dehydrogenase (Xdh) locus in several species of *Drosophila*. Subjecting homogenates to different periods of high temperatures gave evidence of four classes of enzyme. These were either inactivated by five minutes, ten minutes or fifteen minutes of heat, or still active after fifteen minutes. They showed that such enzymes behaved as though they were inherited at the Xdh locus, and all four types showed the same band before heat treatment. Bernstein *et al.* found that populations varied in the number of these iso-alleles that they possessed. One population of *D. montana* showed only two alleles on a basis of electrophoretic mobility, and each of these contained only one thermal category. A population of *D. americana americana*, however, possessed four electrophoretic alleles, but two of these could be divided into two thermally distinguishable alleles each, giving a total of six alleles in all. From their survey of the Xdh locus in several species of *Drosophila*, they found 32 temperature-sensitive alleles where electrophoresis showed only eleven. This suggests there to be 2.9 times as many alleles as they could detect by electrophoresis. We have no idea how general this result may be. But it *must* result in an underestimate of the number of alleles in a population.

Underestimation will also result from the smallness of some samples, for rare alleles may be missed. Indeed some workers deliberately omit alleles with frequencies of less than 5%.

We have at present little idea of the relative importance of these points, although we know that all are valid and must affect the estimates of polymorphism. Despite our ignorance, we must proceed to a consideration of the significance of a level of polymorphism of 33%, for it has important consequences.

Selective neutrality, or otherwise, of the alleles involved

Lewontin and Hubby[176] themselves suggested that many of their protein polymorphisms might be selectively neutral. They were inclined to dismiss this possibility as 'not a satisfactory explanation' on the grounds that some populations were already known to be large from previous research by several workers, including Dobzhansky and Wright.[80] These studies had, in fact, shown that many *D. pseudoobscura* populations were far too large for differentiation by genetic drift alone. Furthermore, Lewontin and Hubby point out that some of their populations were small, and yet equally variable, and in fact the absence of pure local 'races' argued against selective neutrality, although we have already seen that a very low level of migration is sufficient to prevent differentiation at neutral loci.

A more recent report by Avise and Selander[8] casts further light upon this. They examined enzymic and other protein variation at 17 loci in populations of cave-dwelling fishes of the genus *Astyanax* in Mexico. Samples were taken from three cave populations, and six on the surface. In two of the caves, Pachon and Los Salinos, the fish were pale and eyeless, while in the third, Chica, they were more variable in both eye size and colour. The former two samples gave estimates of 0% and 11.7% of polymorphic loci in the populations. The Chica fish gave an estimate of 29%, compared with the surface populations which ranged from 29–40%. Avise and Selander state that water periodically influxes into the Chica Cave, bringing surface fish with it, but that the other two caves are much more isolated. They suggest that since the cave populations are so small, random processes may have resulted in a loss of variability in them, with a consequent increase in homozygosity. They suggest that the considerable morphological specialization in these caves is further evidence of their isolation. It is also, of course, evidence that selection is acting at least at some of the loci in these populations: those controlling anatomical characteristics. Chica fish are much more variable, both enzymically and morphologically. It is not possible to distinguish here between the hybridization of a cave population with recently influxed surface fish, and a cave population in the process of evolving its characteristic features of pale colour, small eyes and homozygosity.

This, then, is a tentative possibility of random genetic drift at protein-coding loci in natural populations. There is the possibility of selection producing homozygosity for a series of alleles in the subterranean populations, but different alleles have become fixed at some loci in Pachon and Los Salinos, which is circumstantial evidence against selection.

An alternative explanation to both of these would be that the number of fish that founded the populations in Pachon and Los Salinos was so small that many alleles were not represented in the parents. In the absence of subsequent immigration or mutation, these populations would remain genetically depauperate. A 'founder effect' similar to this has been described by Prakash *et al.*,[221] who compare three widely separate populations of *D. pseudoobscura* from continental North America. With the exception of a couple of loci that are associated with a particular inversion, and therefore probably subject to a variety of selective forces, there is no evidence of difference in the amount of polymorphism between these populations. Furthermore, several polymorphisms show similar gene-frequencies between these populations, and one appears to be clinally varying.

A population was also studied from Bogota, Colombia, and this was much less polymorphic. At several loci, the only allele to be found was the one that is the commonest in the other populations. Such a pattern would be anticipated if a few individuals migrated or were transported to a new area where they founded a new and completely isolated population. These results suggest strongly that a pattern of gene-frequencies showing a high level of monomorphism need not be evidence for selective neutrality, but may merely reflect an historical accident.

Great similarities in gene-frequency between widely separate populations, as found by Prakash *et al.* in the North American *D. pseudoobscura*, were also found by Ayala *et al.*[10] in tropical american samples of *D. willistoni*. They suggested that this uniformity was not due to balancing selection, but rather that migration in *D. willistoni* is so great that its populations are effectively continuous. Now Kimura and Crow[153] defined a parameter which they termed the 'effective number of alleles' in a population. This is basically the number of selectively neutral alleles, all of equal frequency, that would give rise to the amount of heterozygosis measured in nature, and is given by

$$n_e = 1 + 4N_e v$$

where N_e is the effective population size, and v is the mean mutation rate between alleles. If *D. willistoni* has an effective population size of a thousand million, with a mutation rate of 10^{-7} at enzyme loci, Ayala *et al.*[10] pointed out that the effective number of alleles per locus would be about 400. They calculated that it was, in fact, only 1·4 and suggested that the discrepancy must mean that selection, and not mutation and drift, was controlling the polymorphism.

Ohta and Kimura[208] then revised the theoretical work of Kimura and Crow [153] to allow for the fact that several structural alleles might give rise to the same electrophoretic band, as Hubby and Throckmorton[127] have subsequently shown to be the case. They found that, using a simple model of single step differences in mobility for the substitution of various amino acids, the effective number of alleles came to be

$$n_e = \sqrt{1 + 8 N_e^v}$$

For *D. willistoni*, this now gives a value of $n_e = 28$, which is still about 20 times too high, but Ohta and Kimura suggested that the populations were still evolving, and that, in time, such a figure might be attained.

In their computations, Ohta and Kimura[208] had assumed a potentially infinite number of alleles, but King[157] has now explored the consequences of allowing for only a finite number of electrophoretically distinguishable alleles. He concludes that the number of such variants in a population would be fairly small. Furthermore, although it would increase with population size, there would be a limiting population size beyond which the number would not increase very much. This avoids the otherwise embarrassing finding of Ayala[9] who showed that the pattern of allele frequencies was similar in a series of small island populations of *D. willistoni* from the Caribbean and in the supposedly continuous mainland populations. King showed that this could arise if the polymorphisms were due to mutation and drift, although in fairness, he pointed out that it could also be due to selection or even an interaction between all of these forces.

The use of the effective number of alleles in the problem of enzyme polymorphism has been extended further by Ewens[92] and by Johnson and Feldman.[134] Ewens derived formulae that relate the observed number of alleles in a sample to the effective number of neutral alleles in the population. Johnson and Feldman used this to show that, if the alleles in nature were selectively neutral, the ratio of observed to effective number of alleles should increase rapidly in relation to the observed number, provided that the samples are sufficiently large (>50). However, when they examined enzyme data from a variety of sources they found that the ratio did not increase, quite the reverse, it went down, and the decrease was statistically significant. This suggested that the assumption of the selective neutrality of *all* enzyme polymorphisms might be invalid, and, as we have seen, the results of Johnson[133] and Gillespie and Langley[109] lead us to the same conclusion. Nevertheless, the computations of Lewontin and Hubby,[176] based upon the idea of heterozygous advantage at every locus, lead to equally embarrassing results. They referred to the dilemma that the level of polymorphism is too high to be explained other than by selection, and yet their model of selection suggested that the level was so high as to impose an intolerable reproductive burden upon the populations.

Alternative models based upon heterozygous advantage

The evidence that selection was pertinent to the amount of heterozygosity, together with the dilemma that stemmed from it, gave rise to a series of thoughtful papers. These discussed alternative models of selection to the purely multiplicative one constructed by Lewontin and Hubby. There is no doubt that under certain circumstances multiplicative models may be correct. If selection acts upon two loci at different times during the life of an organism the mortality associated with one locus will be quite independent of that relating to the other. For example, the moth *Lasiocampa quercus* is a moorland insect inhabiting northern Britain. In the adult form, the males are typically reddish-brown, and the females are pale yellow. A recessive gene is present in some populations that confers a greenish-black pigment upon the adults.[102] This form appears to be a standard industrial melanic, for it is found in increasing frequencies in manufacturing districts of northern England. The larvae of this species are normally brown, but again there is a recessive gene which acts to make the larva black when homozygous. These two genes are at separate loci, and visual selection for crypsis will act upon them independently, as follows. The larval gene manifests itself first, and excess numbers of the more conspicuous phenotype will be eaten. When the animals are in their pupae, different loci are involved in the production of pigment, and after emergence selection now acts upon these via the adult phenotype.

We can see the effects of such a system upon the ecological genetics of *L. quercus* by recourse to Table 4.4. There are two larval and two adult phenotypes, so an individual will fall into one of four classes. Let us suppose that predators remove one quarter of the melanic larvae in an industrial area, and half of the typicals. Suppose, also, that similar mortality levels operate against the surviving adult. We can see from Table 4.4 that, of those zygotes which have the genetic constitution to be non-melanic as both larvae and adults, one half will perish before pupation, and half of the survivors after emergence, but before mating. Thus the selective mortality is three-quarters, and the fitnesses are multiplicative.

Table 4.4 Survival of *Lasiocampa quercus* under an hypothetical selection regime. In an urban environment, half of the available typicals and one quarter of available melanics are eaten. This assumes the same predation levels acting upon larvae and adults. Note the multiplicative fitnesses.

Larval phenotype	Adult phenotype	Proportion surviving to pupation	Proportion of these surviving to mate	Total survival
Melanic	Melanic	$\frac{3}{4}$	$\frac{3}{4}$	$\frac{3}{4}$ of $\frac{3}{4} = \frac{9}{16}$
Melanic	Typical	$\frac{3}{4}$	$\frac{1}{2}$	$\frac{1}{2}$ of $\frac{3}{4} = \frac{3}{8}$
Typical	Melanic	$\frac{1}{2}$	$\frac{3}{4}$	$\frac{3}{4}$ of $\frac{1}{2} = \frac{3}{8}$
Typical	Typical	$\frac{1}{2}$	$\frac{1}{2}$	$\frac{1}{2}$ of $\frac{1}{2} = \frac{1}{4}$

In parenthesis, we might note that in an industrial environment where both melanic alleles are segregating, the most successful animals will express the melanic phenotype at both loci. Linkage and consequent linkage disequilibrium (see p. 144) would be favoured, and predators would produce an excess of the larval melanic-adult melanic combination. This appears to be the case in one population, since Ford[102] reports that although the adult melanic is found at a frequency of 4%, about half of the rare melanic larvae produce this adult phenotype. Perhaps here is a situation where linkage disequilibrium is evolving within a contemporary population.

Multiplicative interactions in fitness as exemplified by *Lasiocampa quercus* may be quite widespread. Nevertheless, they are not always the case, and alternative models of selective interaction were examined.

Milkman[187] suggested that it was unreal to regard all loci as operating independently, and consequently the multiplicative model of fitness should not be used. Furthermore, it seemed unreasonable to him to propose that a population composed entirely of homozygotes should have a fitness of $(0.98)^{2000} = 2.8 \times 10^{-18}$, and consequently be almost unable to survive. Indeed, the cave-fishes of Avise and Selander[8] seem to bear out this point.

Sved, Reed and Bodmer[252] also raised these points in a detailed appraisal of the problem. They suggested that one of the difficulties arose because fitnesses were related to an optimum genotype, even though this genotype may be absent from the populations under consideration. In fact, if each of 2000 polymorphic loci in *D. pseudoobscura* carries two alleles that are equal in frequency, the all-heterozygote genotype will have an unbelievably low frequency of $(0.5)^{2000}$, which is less than 10^{-600}. Sved *et al.* proposed that the fitness of each genotype should be related to the *mean fitness of the population*, and not to an arbitrary, and almost never occurring standard.

We can now adapt their computations to this problem of 2000 segregating loci in a population of *D. pseudoobscura*, and continue with the assumption that a homozygous individual has a fitness that is 2% less than a heterozygote. The average individual will be heterozygous at 1000 loci, and will have a fitness of $0.98^{1000} = 1.7 \times 10^{-9}$. We have already seen that the mean fitness of the population is 1.9×10^{-9}, so under Sved's model, the fitness of this average individual will be $1.7 \times 10^{-9}/1.9 \times 10^{-9}$ or 0.903 relative to the population. The relationship between the number of heterozygous

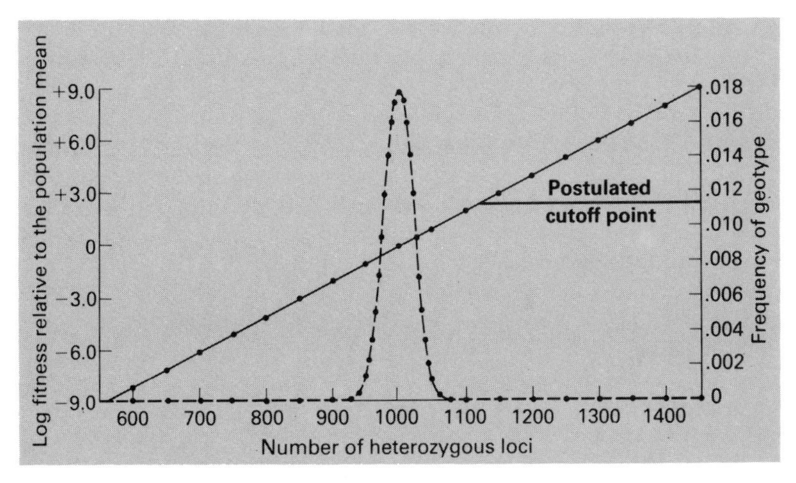

Fig. 4.5 A graphical representation of the model of Sved *et al.*[252] applied to the problem of 2000 loci each with two equally frequent alleles. The solid line shows the relationship between the fitness of an individual (relative to the population mean) and the number of loci at which it is heterozygous. The broken line shows the frequency of each class of individual. Note that the majority of individuals possess between 950 and 1050 heterozygous loci, and that they have fitnesses close to the population mean. Sved *et al.* suggest that no individual should have a fitness more than ten times the population mean, and suggest an upper asymptote as shown. Individuals affected by this arbitrary reduction of fitness comprise less than 0·0001% of the population.

loci and this relative fitness is shown in Fig. 4.5 There is still a clear superiority of individuals with a large number of such loci, and Sved *et al.* suggested that this be reduced by invoking a threshold. They argued that it is unlikely that the fittest individual is ever more than ten times as fit as the average, and so an upper limit could be set beyond which additional heterozygosity will not result in an increase in fitness. It is straightforward to determine how many heterozygous loci result in a fitness that is ten times the population mean by solving:

$$\frac{0{\cdot}98^{(2000-x)}}{0{\cdot}99^{2000}} = 10$$

The answer is 1120, and so Sved *et al.* propose that all individuals with more than that number of heterozygous loci have the same fitness, as represented by the broken line in Fig. 4.6. They further suggest smoothing the angularity in the relationship to make the model slightly more realistic. Imposing a threshold in this way has virtually no effect upon the mean fitness of the population because (as the graph shows) individuals that are heterozygous at more than 768 are so rare.

Sved, Reed and Bodmer[252] support this model with evidence from studies on inbreeding depression. The loss in reproductive vigour of a population when it has been subjected to several generations of inbreeding will be discussed later. Sved *et al.* showed, however, that the decline in performance would be much more rapid than experimentation indicates if the maximally heterozygous individual were many times more fit than the average. They believe that their computations on inbreeding show that many of the segregating loci could not be maintained by multiplicative selection for the heterozygotes.

King[155] also suggests escaping from the dilemma by assuming that the worst individuals are removed by selection up to a threshold number of survivors. The position of the cut-off point would depend upon the ecological conditions, and would operate downwards. Thus individuals survive if there is ecological space for them, and this depends upon the ability of those competitors which are actually present in the population, and again not to some absolute standard genotype.

These models have introduced a note of realism into the problem. They have not, however, solved it. A perusal of the literature shows that alternative and more complex systems are proposed. These are outside the scope of a text such as this but essentially consider the effects of the linkage of loci into larger units. Perhaps the most important is that due to Franklin and Lewontin.[104]

FREQUENCY-DEPENDENT SELECTION

So far we have assumed that the fitness ascribed to a genotype is predetermined and unvarying. There is no reason why this should be the case. Indeed, it is often biologically unlikely that the fitness of a particular genotype should remain constant whatever the frequency of the class which it occupies within a population. There has been a considerable upsurge of interest in recent years towards the questions of whether selective values vary, and, if they do, how they vary. The initial simplest starting point is to consider the effects of relating fitness directly to the frequency of the gene or, perhaps more realistically, of the genotype.

R. A. Fisher[96] seems to have been the first person to state that a balanced polymorphism may follow if the fitness of a genotype is related to its frequency. It is intuitively apparent that this may be so. If the fitness of a genotype increases as its frequency in a population declines, there will be an increased proportion of the alleles composing the genotype in the next generation. Conversely, if the fitness declines as the frequency increases, the trend will be reversed.

General and rigorous mathematical proofs of the stability or otherwise of the polymorphisms which are maintained are not yet available, although several restricted models have been produced, such as those of Haldane and Jayaker,[121] Clarke and O'Donald[48] and Cockerham *et al.*[55] We shall not consider these at all. Most of the models give intuitively obvious equilibria, and their stabilities need not concern us. Of more importance is the way in which frequency-dependent selection can act in a population, and its significance to the problem of genetic polymorphism. We will consider in turn several ways in which frequency-dependent selection can operate, considering whether these can be related to the problem of enzyme and protein polymorphism that we have been discussing during much of this chapter. The three broad categories into which we shall divide frequency-dependent selection are (i) competition between genotypes for limiting substances; (ii) mating differences; and (iii) interactions between species, especially predator-prey relationships.

Competition

When two categories of organisms, be they species, genotypes or sexes, have slightly different ecological requirements, there will be competition between individuals in the same class. Should one of the requirements be in short supply the competition will be more intense in the category requiring it, with the possibility of reduced viability ensuing. This has been shown by Sokal and Karten[247] using *Tribolium castaneum*, Petit[214] and Anxolabehere and Periquet[5] with *D. melanogaster*, and Harding [122–3] using different species of plants.

The studies of Sokal and his colleagues were among the earliest and show how complex the effects can become. Sokal and Karten[247] used the flour beetle *Tribolium castaneum* to study competition between individuals carrying the mutant 'sooty' gene and those carrying the wild type. The sooty locus was used because heterozygotes could easily be distinguished. Sokal and Karten counted out the eggs of sooty, heterozygous and wild type beetles in Hardy-Weinberg proportions, and placed them in pre-determined quantities of flour. They then counted the number of emerging adult beetles of each genotype, and as might be expected, this depended upon the density of eggs in the flour. Their results are complicated by the fact that there is a relationship between the survival of individual genotypes and the egg density, irrespective of genotype frequency. The wild-types performed well at all densities. The heterozygote was superior to the sooty homozygote at low density, but this relationship was reversed as the density increased.

The influence of density was eliminated by statistical manipulation of the data, and it then became apparent that there was good evidence of a relationship between the performance of the genotypes and the gene-frequency of the population of eggs. Sooty beetles survived better than heterozygotes when the frequency of the sooty gene was 0.25, but this was reversed at frequencies of 0.5 and 0.75. The wild-type was less successful than the heterozygotes when the sooty gene frequency was 0.75, but at the other two frequencies the heterozygotes were less successful.

These results seem to indicate that the relative survival, and presumably the competitive ability, of the three genotypes at the sooty locus are affected by the frequency of genes or genotypes in the experimental populations. It is possible, indeed probable, that these effects are not due to the sooty locus itself, but to alleles of other genes linked to it. The differences detected by Sokal and Karten are then due to variations in the performance of the strains of beetles which are marked by the genes at this locus.

There are two possible reasons for the results. Firstly, it could be that each of the competing strains is receiving a different, vital, but limited, resource from the vial of flour. A larger number of one genotype would then result in greater competition between those individuals, with a consequent reduction in success compared with a lower density. Alternatively, some metabolite could be produced by each genotype which adversely affects competitors of the same genotype. It is difficult to distinguish between these in the absence of closely defined food media which are analysed before and after each experiment.

A series of experiments on *Drosophila* by Dawood and Strickberger[67] argue in favour of the addition of biotic residues to the medium. They allowed larvae of a particular strain to feed in vials of medium for two days before killing them by freezing. They then added new larvae of the same or a different strain and compared the number of adults that emerged. Some combinations of 'conditioning' strain and 'experimental' strain resulted in an increased survival of the latter, others in decreases. Enhanced survival is difficult to explain in terms of the removal of vital nutrients, and Dawood and Strickberger argue in terms of the addition of 'biotic residues'. There can be little doubt that the churning of the medium by larvae for two days will result in an increased growth of yeast, which might favour some strains more than others. It is known from other studies that strains differ in their behaviour, both as larvae and adults, so differential amounts of churning might occur, which in turn may result in interactions of the type observed.

Kojima and Yarbrough[164] have examined the possibility that frequency-dependent selection acts at the esterase-6 locus of *Drosophila melanogaster*. There are two alleles here, called 'fast' and 'slow' from their electrophoretic mobilities. Kojima and Yarbrough derived a series of lines from a population which had been established in the laboratory for 30 generations as a large randomly mating unit. These lines were then mated in a controlled random fashion to produce crosses of four kinds: FF × FF, FF × SS, SS × FF and SS × SS (female first). The females then carried eggs of known genotypes and were set up in culture in Hardy-Weinberg proportions corresponding to different gene-frequencies. They were allowed to lay eggs for three days, and adult males were assayed on emergence for their esterase genotype. The viability of the genotypes in each culture was determined by dividing the proportion of emergents that occurred by the proportion expected from that culture, allowing for known differences in the fecundity of the females. These viability estimates are shown in Table 4.5, and it can be seen that they are frequency-dependent. The viability of the SS genotype rises as the frequency of S declines, and that of FF is similarly related to the frequency of the F gene.

Similar results have also been reported by Kojima and Tobari[163] using the alcohol dehydrogenase locus of the same species. Criticism can be levelled against both of these studies since the populations from which the lines were derived had not been in the laboratory very long, and had been founded by the mixing of two inbred stocks. There might be considerable non-random associations of alleles around the enzyme locus under experimentation. If any of these alleles affect viability, the results which Kojima and his colleagues report could be due to these

Table 4.5 Adult to adult viability estimates for three genotypes of the esterase-6 locus in *D. melanogaster*. The fecundity of pre-fertilized females was checked, and they were placed in vials in such proportions that populations of eggs would be laid at the gene frequencies in the margin, and in Hardy-Weinberg equilibrium. By comparing the productivity of adults with the predicted outcome, viability estimates were produced. These gave *prima facie* evidence of frequency-dependent selection, for the viability of a genotype was higher when it was present in the population at a low frequency. From Kojima and Yarbrough.[164]

Frequency of F	FF	FS	SS
0·7	0·759	1·140	1·713
0·5	0·842	1·034	1·099
0·3	1·677	1·726	0·697

alleles rather than the enzymes.

A major study by Yamazaki[281] using the sex-linked esterase-5 locus of *D. pseudoobscura* gave no evidence of frequency-dependence. Nor did he find evidence of conditioning effects by one genotype upon the performance of experimental genotypes. However, Kojima and Huang[162] found that *D. melanogaster* appeared to suffer reduction in viability when reared on medium that had been conditioned by the same esterase-6 genotype. Clearly, more experiments are needed along these lines to determine the generality or otherwise of frequency-dependent competition. When these results are reported, whether positive or negative, with precisely controlled individuals being used, it will become possible to pass judgement.

Frequency-dependent mating performance

This is well-established, both in animals and plants, though it is by no means a general phenomenon. It is known in plants through the self-incompatibility system, and from mating-behaviour studies in several groups of animals.

In the case of the former, there is a genetic system among some genera of plants which prevents, or at least reduces, self-fertilization. In some species, a series of alleles segregate at a single locus, and pollen will not grow successfully on the stigma of a plant carrying the same gene. This prevents self-fertilization, and so reduces the possibility of rare deleterious recessive genes coming together to produce disadvantageous homozygotes. Any system which acts to prevent this will probably be selectively advantageous, for a wastage of gametes may result from the production of such zygotes. At least three alleles are needed for this system to work, however, for, if fewer are present, homozygotes will only be able to fertilize the alternative homozygote with the consequent production of unfertilizable heterozygotes.

Individuals carrying a rare incompatibility allele must be at a considerable reproductive advantage under a self-incompatibility system. They will produce pollen that carries this rare gene, and will be able to fertilize virtually every other plant in the population, causing the allele to increase in frequency. Pollen bearing a commoner gene will be able to fertilize rather few plants, so the fitness of the individuals producing it will be correspondingly reduced.

Emerson[87] first described the genetics of a system such as this when he analysed the self-sterility system of *Oenothera organensis*. Lewis[173] later recorded 45 alleles at the relevant locus in this species which is restricted to a small mountainous area of New Mexico. The total population size of this species was estimated at no more than 500 individuals. It is unheard of for a population to carry such a large number of alleles at a locus when its size is so low: random processes will result in the rapid loss of many of them. The high number was variously attributed to a high mutation rate and small, sub-structured populations, although Ewens[91] showed that it could be explained if the rate of loss due to random effects were drastically reduced. The self-sterility system of *O. organensis* produces a strong advantage to rare alleles, and a consequent increase towards an equilibrium position where all the alleles are (presumably) equal in frequency.

The phenomenon where individuals mate with dissimilar members of the population more frequently than would be expected by chance is termed dis-assortative mating. The reverse process is assortative mating, and both are to be found widely among animals and plants. An early example of dis-assortative mating in animals comes from studies of Sheppard[238] and Sheppard and Cook[240] on *Panaxia dominula* from Cothill Fen. As we described in Chapter 2, this species is polymorphic for the *medionigra* gene in that population. There are three distinguishable phenotypes, and Sheppard and Cook established a series of trios of moths, with two virgin individuals of one sex and one of the other. The two individuals were phenotypically distinct, so the singleton sex had a choice of mate. They recorded the first mating in a variety of trios with the results shown in Table 4.6. When confronted with a choice between a *dominula* and a *medionigra* mate, *dominula* individuals of both sexes chose a greater number of the dissimilar form. The *medionigra* individuals were somewhat less selective, but dissimilar forms were still chosen more often than would be expected by chance.

There is no evidence that differential mating occurs in the wild, but

Table 4.6 The results of a series of mating preference experiments using *Panaxia dominula*. The data have been pooled from the results given by Sheppard and Cook.[240] In each experiment a single individual of one sex was given a choice of two mates: one of similar and one of dissimilar phenotype.

The results of the male choice experiments are not quite significantly different from random ($p > 0{\cdot}05$), but the females show a highly significant preference for the unlike phenotype ($p = 0{\cdot}001$).

Experiment	Phenotype chosen: like	Phenotype chosen: unlike	Comparison with 1:1
Male choice	34	52	$\chi^2_{(1)} = 3{\cdot}8$
Female choice	39	74	$\chi^2_{(1)} = 10{\cdot}8$

if it does, then when a phenotype is low in frequency it will be at a reproductive advantage. This will be particularly so if *P. dominula* mates several times. Whenever an individual of the common phenotype is faced with a choice of mates, it will preferentially choose the dissimilar, and rarer, form. This will act to increase the frequency of the rare alleles and the frequency of heterozygotes in the population.

Dis-assortative mating has been demonstrated in the wild by Lowther[178] in his studies on the variation of the white-throated sparrow (*Zonotrichia albicollis*). This species shows discontinuous variation in the colour of the crown stripe, which is under simple genetic control. He recorded the phenotypes of 110 mated pairs from a park in Ontario, Canada, and obtained the results shown in Table 4.7.

Table 4.7 The composition of 110 mated pairs of white-throated sparrows from Ontario, Canada. (From Cowther.[178])

Colour of male's crown stripe	Colour of female's crown stripe: Tan	Colour of female's crown stripe: White
Tan	4	27
White	79	0

Assortative mating is not usually as striking as this, however, although Levin and Kerster[171] have provided a fine example in plants for a quantitative character. *Lythrum salicaria* is a common plant in the eastern U.S.A. and Britain, and is usually pollinated by insects. It is very variable in height, and Levin and Kerster have shown that, in the

American locality studied (Indiana), adjacent plants differ by an average of 11.9 cm. The principal pollinators that they studied were honey-bees, and they argued that if these insects were moving at random between plants the distance moved up or down from one plant to the next should have been of similar size. The mean vertical displacement of the bees was, however, only 6.4 cm, which is statistically less. It seems that the bees are remaining at a fairly constant height, and this suggests that plants of similar height will be visited on consecutive pollinating visits. They confirmed this by showing that the bees tended to visit one or other of two strains of *L. salicaria* which were intermingled. These strains were uniform in all characters apart from height. When the taller plants were trimmed back, the pollinators displayed no preference which suggests that height alone was responsible for the limitations of inter-strain flights.

Assortative and dis-assortative mating have different effects in their interactions with the maintenance of polymorphisms. When unlike individuals mate more frequently than would be expected, heterozygotes are produced in relative excess, and this acts to maintain the polymorphism. Conversely, if similar phenotypes mate more often, there is a deficiency of heterozygotes, and the polymorphism is unstable.

There seem to be no published studies on the possibility of non-random mating involving enzyme or protein loci. It is known that there are differences in reproductive success between individual *Drosophila* whose inversion patterns differ. Brncic[18] introduced male *D. pavani* to virgin females and examined the karyotype of successful and unsuccessful females. He found a greater frequency of heterokaryotypes among the successful females and suggested that this might be due to an enhanced sexual activity associated with heterosis. Other differences in mating preference between strains, or inversion types of *Drosophila* have been shown which cannot be ascribed directly to the karyotype. There are several studies which indicate that, in a competitive situation, rare males gain a reproductive advantage by being accepted more frequently in copulation. This has been shown by Petit,[211–13] Ehrman,[81–3] and Spiess,[249] among others, and most are agreed that the reverse situation is less frequent and less marked. Rare females do not gain such an advantage. This is not too surprising, for Bastock and Manning[11] showed that female *D. melanogaster* are more discriminating than males, and usually do the rejecting of a sexual partner.

Ehrman[84] established that the females can distinguish between rare and common males by their scent. She used an observation cage in which she placed five pairs of one karyotype and 20 pairs of another.

The wings were clipped to aid in identification of the karyotypes, and this was reversed between experiments as a control. Mating pairs were noted every six minutes for several hours. The rare males were involved significantly more often than would be expected on the basis of their frequency. This result is fairly standard for such an experiment, as is her observation that there was occasionally a rare female advantage. Ehrman then fastened a second observation cage below the first and placed 15 pairs of the rare karyotype in it, separated from the experimental flies by a layer of cheese-cloth. The advantage of the rare males then disappeared completely (Table 4.8) suggesting scent, sound or physical contact through the mesh to be important. She eliminated the last of these by separating the two cages with two sheets of cheese-cloth, leaving one centimetre between them. Direct contact was no longer possible, but the flies still mated in proportion to their frequency.

She provided strong circumstantial evidence that scent rather than sound was important, by gently blowing a current of air between two cages in which the same number of flies were present as before. When the current blows from the mixed population towards the 15 pairs of flies of the same genotype as the experimentally rare flies, there is an

Table 4.8 The number of arrowhead (AR) and chiricahua (CH) males recorded mating in a series of experiments using a two-chambered cage. The experimental cage contained males and females in the proportions reported; the second cage includes an additional 15 pairs of the rare forms. When air is blown from the experimental cage towards the second, mating departs significantly from random (indicating an advantage to the rare males). When the air current is reversed, matings are random, which suggests that the odour of the males in the second cage is overwhelming the advantage accruing from rarity in the experimental cage. From Ehrman.[84]

Number of pairs							
Experimental cage		Second cage		Air current direction	Number of males mating		
CH	AR	CH	AR		CH	AR	χ^2
20	5	0	15	→	45	58	36·13*
20	5	0	15	←	27	83	1·42
5	20	15	0	→	62	44	30·65*
5	20	15	0	←	88	19	0·34

* Indicates a highly significant departure from random mating when the air current blows from the experimental to the second cage.

advantage to the rare males, which disappears if the current flow is reversed. This implies that the females are unable to detect that some males are rare in their immediate vicinity when air is blown in to them from a cage that contains more of these rare flies. Perhaps not surprisingly, this aspect of the mating behaviour of *Drosophila* has not been studied in the wild. We can see where its evolutionary advantage lies, however. Many species of *Drosophila* breed in small habitats: rotting fruit, fungi, flowers, etc. In these situations, flies from a single site are likely to be related and carry similar food odours. Remembering again that most individuals carry deleterious recessive genes, sib-mating may result in a reduction of vigour. Consequently, choosing mates who smell differently from the mass of the population will increase outcrossing and reduce the chances of inbreeding.

Predator-prey relationships

Phlox drummondii has two cultivars called 'Twinkle' and 'Nana compacta' which differ only in the structure of the corolla lobes. The former has laciniate lobes and the latter's are complete. They apparently are freely self- and inter-compatible and will grow well in mixed stands. Levin[170] grew these two plants randomly in a grid pattern, and observed that both were visited regularly by insect pollinators, especially butterflies. The ratio of the two varieties differed in five experimental plots, and Levin collected seed from 'Nana compacta' regularly through the fruiting season. From the relative number of hybrids and 'Nana compacta' growing from samples of the seed he was able to estimate the frequency of outcrossing, as shown in Table 4.9. There is no difference between the three intermediate ratios, but outcrossing is significantly reduced at the extremes. Levin also showed that outcrossed seed is usually fitter, for there are problems of inbreeding resulting from intra-strain fertilization. So, here is a situation where the rare form is at a disadvantage, for less of its seed is outcrossed and hence the plant has a lower fitness. This is in contrast with the rare form advantage observed in *Drosophila* mating cages, and inferred from studies on *Panaxia dominula*. The reasons for the difference are not hard to see. The phenomenon stems not from the plant itself, but from the animals that are responsible for its pollination. In fact, this method of pollination is more closely related to predator behaviour, for the pollen vector comes to the plant for food.

Lukas Tinbergen[255] first commented upon the 'searching image' concept in predator behaviour. He and his colleagues found that tits feeding their young on woodland invertebrates tended to bring in high frequencies of particularly common species of prey. He suggested that

Table 4.9 The frequency of outcrossing in various mixtures of *Phlox drummondi* cultivars. Outcrossing is broadly comparable when the cultivars are in reasonable equality. When one or other is at an extreme frequency, the amount of outcrossing declines significantly. This results in fewer pollinator visits to the rarer form and its consequent reduction in fitness, and decline in frequency.

Ratio of cultivars	Number of progeny scored	Frequency of outcrossing
9:1	1432	42·8
3:1	912	57·0
1:1	1085	57·2
1:3	628	61·8
1:9	418	49·4

the successful capture of a certain food species inclines the predator towards further search for the same prey. This results in a greater efficiency of hunting because the predator learns where and how to search, rather than randomly hunting for any food type. Greater efficiency can reduce the energy expenditure of the parent and increase the frequency of feeding the young. It will consequently have survival value.

Clarke[39] argued that this behaviour might be important in situations where a predator was hunting for a polymorphic prey. Success in capturing one phenotype might result in the predator remembering and searching for a similar form next time. Subsequent successes will harden this behaviour. Statistically, the predator is more likely to encounter the commoner phenotypes, and so search images will be established for these forms. The rarer phenotypes might then gain a selective advantage, stemming from their rarity and dissimilarity. Clarke coined the term 'apostatic' selection to describe the phenomenon, from the word apostate used for one who stands out from the ecclesiastical norm in religious belief.

At the time he proposed this, Clarke could find little evidence in support of such a selective system. An early experiment had been performed by Reighard[224] who artificially coloured small fish, and fed them to grey snappers (*Lutionus griseus*). He trained the snappers to take either blue or red prey, and then offered them a choice of both. The fish clearly ate the blue prey before the red when they had been trained upon the former.

Similar results were obtained by Allen and Clarke[1] who fed pieces of coloured paste to wild passerine birds. When the birds were trained to take brown or green baits, they took an excess of that colour when confronted with a mixture. This differential selection could be quite marked, with birds taking the training colour even when its frequency in the experimental choice situation was less than 10%.

These results confirm that birds will develop a specific search image, and that this is strong and effective even when alternative food supplies are readily available. Indeed, Soane and Clarke[246] have shown that searching images are not restricted to predators that hunt by sight. They showed that rodents could be trained to take a particular scented food, and that they took this form preferentially in a choice situation.

Training animals experimentally is not apostatic selection, however, and the crucial experiments concern whether search images can be developed in situations where a predator is confronted with a polymorphic prey. Such an experiment had been performed by Popham, and was reanalysed by Clarke.[39] Popham[216–17] fed different coloured corixid bugs to fresh-water fish. The evidence is strong that the fish took greater proportions of a phenotype when it was common than when it was rare, irrespective of whether it was cryptic or not (See Fig. 4.6). This result has been confirmed by Miller's laboratory experiments with Japanese quail. Some of her results are reported by Manly, Miller and Cook,[180] who also discuss methods of analysing selective predation experiments. She fed the

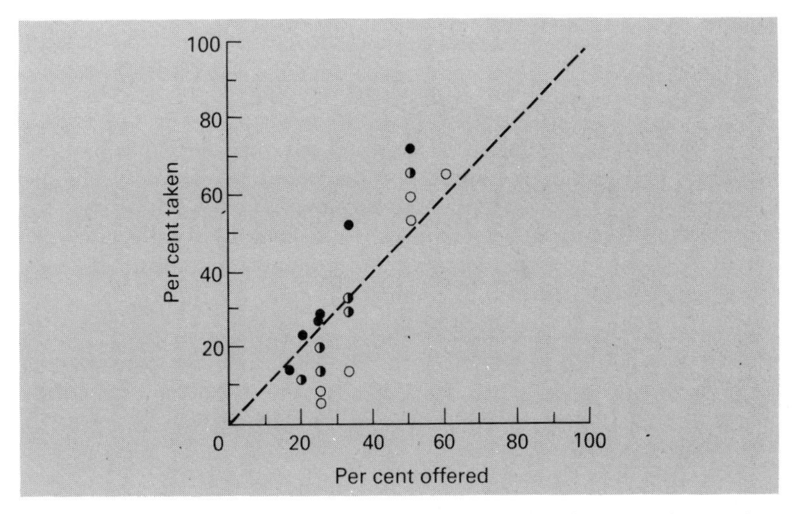

Fig. 4.6 The predation by rudd, *Scardinius eryophthalmus*, on three colour forms of the water bug *Sigara distincta.* ○ denote the most cryptic form, ● the least cryptic. All forms are over-predated when they are common, but are under-predated when they are rare. (Data from Popham.[216–7])

quail with blue and red pastry baits, and her results show that, despite a strong preference for red, the birds took disproportionately more of whichever form was common.

When these experimental results are extrapolated into the wild, it becomes apparent that rare phenotypes of a prey species probably derive a real advantage from their very rarity. This will in turn act to maintain a frequency-dependent polymorphism. There will presumably be an equilibrium position that depends upon the frequency of the alleles, the relative visibility of the phenotypes and the behaviour of the predators. Such a polymorphism has not been verified in nature, however.

Clarke[41] has argued that there is a negative relationship between the abundance of yellow *Cepaea nemoralis* and unbanded yellow *Cepaea hortensis* in some habitats in some areas of Britain. He believes that this might stem from apostatic selection acting upon the two species. If a population of *C. nemoralis* has a high frequency of the yellow phenotype, for reasons unconnected with its appearance, predators will develop search images for it. Consequently, phenotypes of *C. hortensis* will be at an advantage if they differ in colour or pattern. Experimental verification of this by artificially altering the phenotype frequencies of one species but not the other is not yet available. It would be a worthwhile task, for frequency-dependent selection of this form is a powerful potential force in the maintenance of visual polymorphisms.

Heterozygous advantage and frequency-dependent selection are conceptually the two simplest methods for maintaining segregating alleles within a population. They are not the only mechanisms, however. Selection depending upon the population density can also act in this way.[36, 44–5] More complex systems such as selection acting in different directions in the two sexes, or changing its effects upon a genotype with time, will also maintain polymorphism, but these are essentially special cases and do not need to concern us here.

5

The Effects of Selection on Quantitative Characters

INTRODUCTION

The chapters so far have discussed the behaviour of genes at single loci, both in theory and in nature. There is, however, a large amount of genetically controlled variation which is, at first sight, much more complicated than this. Almost every feature of a species shows continuous variation for quantitative characters such as size, or weight, seed number or mating speed. Characters like these are measured in scalar units, rather than scored directly into discrete classes that bear a fairly simple relationship to the controlling loci. The study of these metric traits and their inheritance is covered by the field of quantitative genetics, and is very pertinent to the theory of evolution by gene substitution.

Breeding experiments rapidly show that the inheritance of continuous, or metric, characters is rarely as simple, for example, as the flower colour of peas or the eye colour of *Drosophila*. Offspring are frequently intermediate between their parents, rather than resembling one or the other, as is often the case with more simple genetic situations. Further analysis usually suggests that they are controlled by many loci, each of which has two or more alleles segregating. The term 'polygenes' was coined by Mather[181] to describe these genes, which are individually small in effect, but cumulatively control the magnitude of the quantitative character. It now seems likely that these genes are no different from the major genes of large effects, with which we have been so far concerned, in their inheritance or behaviour. Indeed, it may be that polygenes for one characteristic are major genes for another.

When we come to describe a metric character in a population, a problem at once arises. We cannot identify the polygene involved, so gene-frequencies cannot be calculated. It is only possible to measure the phenotype of every individual in a sample and assess the underlying population by a mean or average. Closer examination of the measurements usually shows that the majority of the individuals in the sample lie close to the mean, with a symmetrical reduction towards the extremes. Such a pattern is similar to the 'normal distribution' of statistics. Statisticians record the spread of observations around the mean as the variance. This parameter reflects the average deviation of all individuals from the sample mean. A large variance implies that some individuals depart widely from the mean, or that the population is variable. Conversely, a small variance indicates that all the individuals are rather similar.

We can demonstrate how a series of polygenes may produce such a pattern using a simple model. Suppose that seven loci are involved, at each of which two co-dominant alleles are segregating with a frequency in the population of 0.5. Let one of these alleles contribute one unit to the size of the character, and the other contribute two units. The smallest individual to be found in the population will be homozygous for a one-unit gene at every locus, and so will be 14 units in size. The largest individual will be homozygous for two-unit genes, and so will measure 28 units. There will be individuals present in the population of all integer sizes between 14 and 28. An individual that measures 15 units in height can only be produced if there is one, and only one, two-unit allele in its genome. There are fourteen positions at which this allele can be situated, and since the alleles are present at a frequency of 0.5, they are all equally probable. This size class will thus be 14 times as abundant as the smallest size class. Individuals measuring 16 units must have two of the two-unit alleles, and there are $14 \times 13/2$ ways of producing such a genome. Again all are equiprobable, and so individuals measuring 16 units will be 91 times as abundant as the rarest, smallest class. It is possible to proceed all the way to size class 24 in this way, and produce a frequency histogram as shown in Fig. 5.1. The mathematically astute reader may have realised that the pattern of distributions obtained is in fact produced by the binomial expansion of $(x+y)^{14}$, where $x = y = 0.5$.

The variance among the individuals in Fig. 5.1 is entirely genetic, for it depends solely upon segregating genes. Variation can also be due to environmental effects, however. Small fluctuations in food supply, differential availability of water or some vital chemical, differences in crowding, and a host of other factors, can produce heterogeneity in the environment. As a result, one situation may be more, or less, favourable for growth and development than another, and individuals will vary in their response to this. The relationship between genetic and environmental components of variance is important. For example, in a human disease, if the genetic component is low, it may prove possible to reduce the incidence of the disease by modifying the environment. Conversely, however, a disease that has a high genetic component will have to be approached in a different way, perhaps by genetic counselling to reduce the probability of production of affected children.

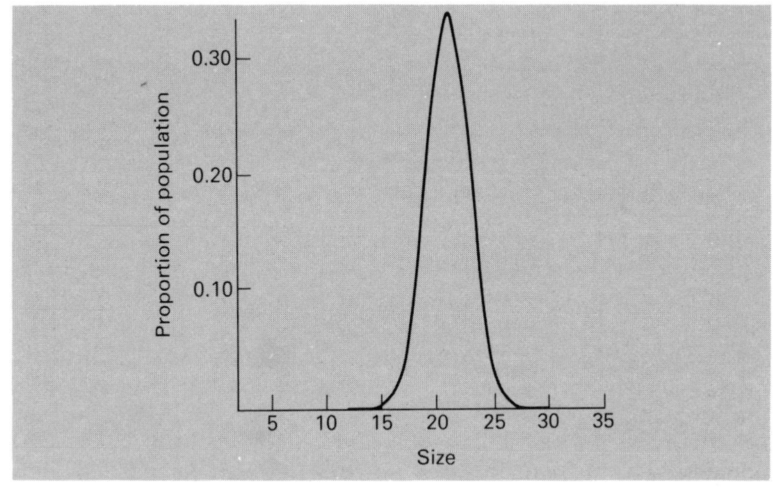

Fig. 5.1 A population distribution generated from a simple model. The horizontal scale shows the size of all the members of a population in which two alleles are segrating at each of seven loci. The vertical scale shows the proportion of individuals in each size class.

It is possible to divide this genetic variance itself into several components. The first, and most obvious, is due to segregating genes within a population. The variance in Fig. 5.1 is of this kind, and is called 'additive genetic variance'. This will occur whenever alleles segregate at a locus and affect the magnitude of the character under consideration.

Further variance can arise from the fact that heterozygotes are not always exactly intermediate between the two homozygotes. In the model discussed above, we assumed that a heterozygote at each of the 7 loci contributed 3 units to the size. This need not be the case. If one gene is dominant, or partly so, then the value of the heterozygote will be closer to one homozygote, and this will cause a change in both the mean and the variance. The component due to this is called the 'dominance variance'.

The third important component results from the fact that the value of a particular homozygote may depend upon the alleles which are present at a quite separate locus. This interaction between genes gives rise to an 'interaction variance'.

The proportion of the genetic variance that is due to additive genetic variance is termed the heritability, and reflects the amount of variation that is due to segregating loci. This parameter is particularly important in agricultural genetics, for, as we shall see, the genetic improvement of a strain depends upon a re-organization of its gene pool following selection. If the heritability is small, then there is little additive genetic variation, and consequently little chance of a selection-based improvement in performance.

SELECTION

Suppose, now, that we have taken a sample and find one individual which lies at an extreme. What can we deduce about it? Firstly, it may have come from a situation where the vagaries of the environment allowed unusual growth and development. Secondly, it may be homozygous at those loci which control the magnitude of the character in question. If the latter be the case, such an individual may be of value in a programme of artificial selection and controlled breeding directed towards altering the genetic, and hence phenotypic, constitution of the population. The remainder of this chapter will be devoted to a discussion of the ways in which selection can be imposed upon metric characters, and the consequences of the selection regimes for the populations involved.

There are three principal patterns of selection that are important in quantitative genetics. These are most easily discussed in terms of simple selection experiments, and are illustrated diagrammatically in Fig. 5.2. The first is stabilizing selection. In this, the individuals which depart most widely from the mean are discarded and only those close to the mean are used as parents in the next generation. The second is directional selection, and in this it is the individuals closest to one extreme which are chosen for future parents. In both stabilizing and directional selection, the ultimate goal is phenotypic uniformity at a desired optimum. The third case is disruptive selection. This is, to a certain extent, the complement of stabilizing selection, for the individuals closest to the mean are discarded, and the extremes are mated together. Perhaps not surprisingly, this results in considerable phenotypic diversity.

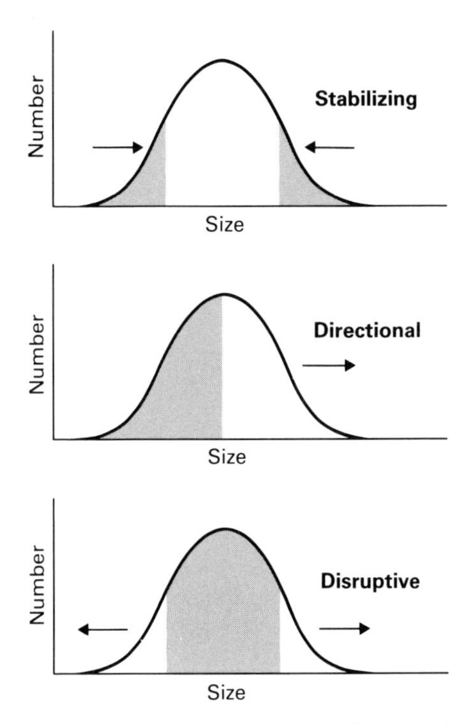

Fig. 5.2 The difference between stabilizing, directional and disruptive selection on the same normal population. The hatched area shows the proportion of the population which fails to survive and reproduce.

These three patterns of selection have different effects upon the population concerned, but they have one property in common. In order that a selection programme can be successful, a sizeable proportion of the offspring are discarded every generation. Consequently, the parents consist of a rather small number of individuals. We have already seen in Chapter 2 that a reduction in the number of parents results in inbreeding, and we must now turn to a consideration of its effects upon quantitative characters.

INBREEDING

It has been estimated that almost every person alive today carries at least one lethal or severely debilitating gene in the heterozygous condition, and there is no reason to suppose that this situation does not extend to most other species as well. Since related individuals have some common ancestry, it is more probable that they will carry the same disadvantageous recessive genes than will unrelated individuals. There is consequently an appreciably greater risk of the offspring of inbreeding being homozygous for some severely debilitating condition.

It is inevitable from the nature of populations that some inbreeding will always occur. For example, in man every individual has two parents, four grandparents, eight great-grandparents, etc. The number of ancestors is related to the number of generations simply as

$$\text{number of ancestors} = 2^{(\text{number of generations})}$$

If we assume 25 years for a generation in man, an individual born in 1966 is separated by 36 generations from his ancestors born about 1066. A child born in 1966 has $2^{36} = 68\,719\,476\,736$ ancestors from the earlier date, which is far larger than the population of the world even today. It is inevitable, therefore, that some individuals will be represented more than once in the family tree: that inbreeding has taken place.

As we have seen in Chapter 2, at the level of the single locus, the effects of inbreeding are fairly straightforward. In a population which has been maintained at an effective size of N_e for t generations, the probability that a single neutral locus in an individual carries two alleles which are identical by descent is related to generation number as follows:

$$F_t = \frac{1}{2N_e} + \left(1 - \frac{1}{2N_e}\right)F_{t-1}$$

It can be shown that the proportion of total loci in an individual that are homozygous is related to generation number in an essentially similar fashion. Furthermore, just as we saw in Fig. 2.2 that the probability of genic identity at a locus increases more rapidly when the population is small, so the rate at which alleles become fixed in a population is also increased if the population is maintained at a low level. Equally importantly, if a series of alleles at a locus are selectively equivalent, the one which eventually becomes fixed is a matter of chance. Just as with random genetic drift and the founder effect, initially more abundant alleles will be more likely to fix, but there is a certain amount of random variation.

A classical study of inbreeding casts further light upon the phenomenon. The reproductive performance of a small population of guinea-pigs subjected to continuous brother-sister mating was studied for 15 years by a series of biologists, and finally analysed by Wright.[273] A whole series of traits associated with fitness declined during the programme: the litter size, the number of litters per year, the percentage of young born alive, birth-weight, etc. These results serve to illustrate a point of considerable importance to quantitative genetics, both theoretically and practically.

In a laboratory breeding programme, animals are usually handled with considerable care, and are cushioned against the adverse effects of disease, hunger, etc. Consequently, the decline in reproductive performance need not adversely affect the fitness of those individuals which are chosen as parents in the next generation. A long programme of inbreeding quite randomly produces identical homozygosis at one locus after another. The fitness of the population steadily declines, but the care with which the individuals are handled ensures that this does not result in extinction of the line. The decline in performance is due to the appearance of genes, in the homozygous state, which were present in the heterozygous condition in the founders of the line, but were concealed by dominance.

This phenomenon is well-known to anyone who has attempted to breed wild-caught animals or plants in the laboratory for several generations. Inbreeding depression almost always reduces the viability of the stock to a large degree in the first few generations. This implies that disadvantageous genes are present in the wild populations, but that they are usually present in the heterozygous state, and are concealed by the dominance of more advantageous alleles. The fact that deleterious, or fitness-reducing, alleles are so often recessive cannot be due entirely to chance. It has led many workers to consider whether the dominance or recessivity of the character itself is a quantitative character that can evolve and change under the influence of selection (see, for example, Sheppard and Ford[241]).

HETEROSIS

When two inbred lines are crossed, there is frequently an improvement in many aspects of the biological performance of the offspring. This is called heterosis or hybrid vigour. Dobzhansky[77] sub-divided heterosis. If it relates to reproductive characters such as the number of offspring he suggested the term 'euheterosis'. When the improvement relates to a more physical parameter, such as back fat thickness in swine, he termed it 'luxuriance'.

Johannsen and Rendel[131] report upon the results of a breeding experiment with pigs which demonstrates euheterosis. They compared litter size, litter weight and early piglet mortality in pure-bred and cross-bred litters, using large white and Swedish landrace pigs.

Some of their results are shown in Table 5.1. The cross-bred pigs are clearly superior in all of these characters. It is interesting to note that in the back-cross of a female to a pure-bred male there is evidence that a cross-bred female produces a larger litter and nurses them better than a pure-bred sow. However, the growth rate drops below that of hybrid piglets as soon as they become independent. It seems that mother pigs can also show euheterosis, in addition to the piglets themselves.

We have seen in the previous section that inbreeding produces a rapid increase in homozygosis. Naturally enough, the crossing of two independently derived inbred stocks gives rise to a great deal of heterozygosity. The increase in performance need not, and usually does not, have anything to do with heterozygous advantage. As we have seen in Chapter 4, this is the situation where individuals which are heterozygous at a particular locus are fitter than their homozygous relatives in that they leave a larger number of offspring to survive and breed. This phenomenon need play no part in heterosis and a simple example will explain why.

Consider a population having twelve polymorphic loci affecting the size of an individual. Let each locus segregate for one dominant and one recessive allele. Suppose that two units of stature are added to every individual for each locus carrying a dominant allele, but only one if the locus is homozygous recessive. If all the alleles have a frequency of 0.5, the mean height of the population will be 21 units. If two lines are established which are maintained under an inbreeding regime without selection, eventually all the loci will become homozygous. Alleles will be fixed randomly, perhaps as follows:

	Loci											
	1	2	3	4	5	6	7	8	9	10	11	12
Founding	A	B	C	D	E	F	G	H	I	J	K	L
population	a	b	c	d	e	f	g	h	i	j	k	l
Line One	a	B	c	d	E	f	G	H	i	j	K	L
	a	B	c	d	E	f	G	H	i	j	K	L
Line Two	A	B	C	d	e	f	g	H	I	j	k	l
	A	B	C	d	e	f	g	H	I	j	k	l

In the first of these lines, loci 2, 5, 7, 8, 11 and 12 are all fixed for the dominant allele, so the size of the individuals is 18 subject to environmental variation. The second line is similarly fixed at loci 1, 2,

Table 5.1 A comparison of litter size, early post-natal mortality, and litter weight in pure-bred and cross-bred litters of piglets at different ages. From Johannson and Rendel.[131]

	Age	Average pure-bred	Average F_1	% increase	Average back-cross	% increase
Litter size	at birth	10·44	10·81	3·5	11·12	6·5
	at 3 weeks	8·47	9·80	14·3	9·16	6·8
	at 8 weeks	8·20	8·71	6·2	8·79	7·2
	at 20 weeks	8·12	8·75	7·8	8·69	7·0
Early post-natal mortality		10·50	3·91	−63·0	7·79	−26·4
Litter weight in kilograms	at birth	14·60	15·56	6·6	14·68	0·5
	at 3 weeks	46·25	50·25	7·8	51·42	10·3
	at 8 weeks	111·16	127·49	14·7	120·04	8·0
	at 20 weeks	377·16	420·90	11·6	405·54	7·5

3, 8 and 9, and the size is 17. If these two lines be crossed, the genetic constitution of the offspring will be:

1	2	3	4	5	6	7	8	9	10	11	12
a	B	c	d	E	f	G	H	i	j	K	L
A	B	C	d	e	f	g	H	I	j	k	l

The offspring will be homozygous for the dominant allele at two loci, for the recessive allele at three, and be heterozygous at the rest. Their size will be 21, which is clearly superior to both of the inbred parental strains. A real situation would probably have environmental effects, and other factors such as linkage, to confuse the picture, but nevertheless we have shown that simple dominance can produce heterosis.

It is most important to appreciate the difference between this situation and heterozygote advantage. Some confusion has arisen in the literature because much of the earlier work on the two phenomena was undertaken using inverted sections of chromosomes. Moos[195] for example, studied various physiological parameters of some chromosomal forms (karyotypes) of *Drosophila pseudoobscura*. He measured, among other variables, pre-adult viability and fecundity which are undoubtedly components of fitness, and growth rate, which is not necessarily so; it is when food supply becomes limiting that the fastest growing individuals may be fittest. Moos found that individuals which were heterozygous for inversions, or heterokaryotypic, had superior performances to homokaryotes. A heterotic situation such as this could easily stem from simple dominance. Suppose that an inversion α includes an advantageous allele *A* and a disadvantageous allele *b* at a separate locus. Suppose also that inversion β contains a disadvantageous, but recessive allele *a* together with the dominant advantageous allele *B*. If there is no recombination between the inversions because their disordered contents prevent chromosomal pairing, the fittest individuals will be the $\alpha\beta$ heterokaryotypes. There is no single locus heterozygous advantage, and perhaps the term heterokaryotype advantage should be used.

We can take this situation further. Simply to observe a superior performance from a heterozygous individual does not allow one to claim that any single locus is manifesting heterozygous advantage. Indeed, it is not even evidence that the locus is involved at all. For example, Hebert, Ward and Gibson[125] have demonstrated natural selection associated with the malic dehydrogenase locus in *Daphnia magna*. This species of crustacean is parthenogenetic during the summer months, and the authors found an increase in the frequency of heterozygous individuals between May and September. They also found that heterozygotes carried more young in the brood pouch than did either homozygote. This situation could have arisen if the population contained two homologous chromosomes that were heterotic in combination. Reproduction over the summer is parthenogenetic and no reorganization of the genetic material takes place. Consequently, all an individual's offspring are identical with each other and the parent. If a chromosomal heterozygote is heterotic, then it will leave more offspring than the homozygotes. Provided that the environment does not change too radically, the heterosis should also be manifest in the offspring, and a further increased production of chromosomal heterozygotes will result. If the two chromosomes carry different malic dehydrogenase alleles, there will be a progressive increase in the proportion of Mdh-heterozygotes. This will have nothing to do with heterozygous advantage at all, for the loci may be selectively neutral, and merely act as chromosomal markers.

We can conclude this brief discussion of heterosis by repeating that merely observing an excess of heterozygotes is not evidence that this form has a superior fitness. It is necessary to prove a functional relationship between the genotypes and the selective agents. A combination of closely linked loci and dominance can frequently give rise to heterosis, and without careful thought one can easily be led astray.

STABILIZING SELECTION

The phenomenon of stabilizing selection was first studied quantitatively by Bumpus.[23–4] He observed the effects of a severe blizzard upon a flock of house sparrows (*Passer domesticus*) near his laboratory at Providence, Rhode Island. He brought a number of weak or moribund birds into the laboratory, and about half subsequently died. Bumpus made museum specimens of all the birds, recording the weight, and various skeletal measurements of each one. He claimed that the sparrows which survived the effects of the cold and snow were less extreme in most of their measurements than those which succumbed. In other words, there seemed to be an optimal size for survival among the birds in the flock: the more an individual deviated from this optimum the less were its chances of survival.

This has been quoted for over half a century as the classic example of stabilizing selection in an animal population. The data have been reanalysed several times as yet more sophisticated statistical techniques became available. In one of the most recent studies, Johnston *et al.*[135] show that there are differences between the sexes in the morphology of the survivors. Adult males can be separated into large and small. The larger males survived better than the small ones, possibly because they were higher in the social hierarchy or 'peck-order'. This would enable them to get more food and so have greater reserves of fat to assist their survival through the period of adverse weather. The small adults survived better than juveniles of the same size, possibly for the same reason, or alternatively because they benefited from the greater experience ensuing from their longer life. On the other hand, the females showed clear evidence of increased mortality among the more extreme members of the population, and Johnston *et al.* consider that the phenomenon of stabilizing selection is justifiably postulation from Bumpus' data pertaining to that sex.

There is an alternative method of seeking evidence for stabilizing selection, which is of greater value to the field biologist. The principle is simple enough: stabilizing selection removes the more extreme individuals from a population and, consequently, the variance of a pre-selection sample should be greater than a sample of post-selection survivors. If we can find a character which can be measured in both adults and juveniles, a change in its variance might be evidence of stabilizing selection. If the two samples are taken from different generations at the same time, we must assume that the environmental effects are the same in the two age-classes.

This method is better for the field biologist who may have difficulty in following a population long enough to observe differential mortality, or who is not as fortunate in his situation as Bumpus. For historical reasons, we will discuss the results of Weldon,[270] who seems to have been the earliest scientist to quantify this kind of study. He examined a population of the land snail *Clausilia laminata* from a mature beech wood in Holstein. He measured various dimensions pertaining to the shape of these snails. Weldon chose his characters intelligently, for the shell of a snail is secreted by glands around the collar of the growing animal, and, once produced, is a permanent record of its growth.

He measured living juveniles after they had grown through a fixed number of whorls of the shell. He could also measure the same dimensions of the adults, for the record of growth is preserved in the shell. He found that the juveniles had a consistently higher variance than the adults. Thus, it seems that the more extreme juveniles fail to survive to become mature adult individuals. This remains an important pioneering study despite two criticisms that we can level at it. Firstly, there is a possibility that Weldon was sampling from an area large in relation to the population size. This might allow heterogeneities in the population history of the different age classes. Secondly, it is possible that the environment was changing, and the selective forces operating upon juveniles may be different from their parents at the same age.

Perhaps the clearest example of stabilizing selection comes from our own species. Karn and Penrose[142] amassed a considerable amount of data concerning birth weights and survival of 13 730 babies born in a London obstetric hospital between 1935 and 1946. Some results of this study are shown in Fig. 5.3. The tinted area of the histogram shows the proportion of babies in each half-pound class from 0.5 lb to 11 lb. It is at once clear that most individuals fall between 5 lb and 9 lb and that the mean size is just over 7 lb.

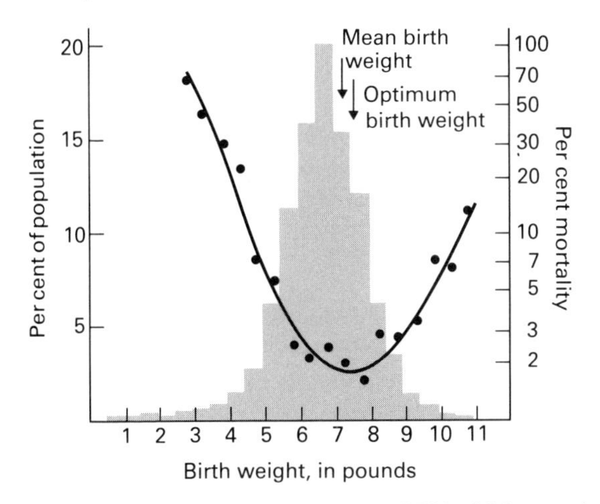

Fig. 5.3 The distribution of birth-weights of 13730 children and their early mortality. The histogram shows the percentage of the sample in each birth-weight class, and the curve shows the percentage mortality. The optimum birth-weight is that associated with the lowest mortality. Data from Karn and Penrose.[142]

Superimposed upon the histogram is the proportion of babies in each size category which did not survive the first four weeks. It is apparent that survival is most probable for babies close to the average weight. Mortality rapidly increases as the birth weight departs from the mean. The optimum size for a baby is that at which it has the greatest probability of survival. This is the birth weight corresponding to the lowest point on the mortality curve. It is interesting to note that this optimum size is 7.5 lb, a clear half pound larger than the mean size.

The reasons for the difference are probably partly connected with the physiological responses of the mother. From the child's point of view, the bigger it is the more likely is its survival. However, large babies have adverse effects upon the mother during birth, and she will tend to produce a child at a safer size for herself.

The causes of death involved in the population of *Clausilia* cannot even be guessed, but the human birth weight is particularly satisfactory because we can fill in some of this information. Small babies are frequently premature, and not fully ready for independent existence. Respiratory, thermoregulatory and digestive systems are not fully developed, and the stresses imposed upon the newborn may be too great. On the other hand, excessively large babies may suffer actual physical damage during birth which affects their morbidity. Alternatively, unusual size may reflect other underlying conditions, which reduce the health and life expectancy of the newborn. In this case the size difference is merely a correlate of some other vitality-reducing factors.

At a purely genetic level, stabilizing selection tends to preserve variability. We have seen that the optimum phenotype is usually an intermediate form, extreme individuals being more or less disadvantageous. Extreme individuals are often more homozygous, and intermediate more heterozygous at loci affecting quantitative characters. Consequently, it is those individuals that are heterozygous which tend to survive; they are phenotypically more uniform, but genetically more diverse.

DIRECTIONAL SELECTION

This selection can be extremely successful in improving the yield of a plant or animal crop. Woodworth, Leng and Jugenheimer[272] report the results of a selection regime designed to modify the oil content of maize kernels (see Fig. 5.4). In 50 generations, the yield of the line which they selected upwards increased three-fold, and there is little sign of a plateau being reached. Their selection downwards is similarly spectacular. The oil content has been reduced from over 4% to only 1%, again with only slight evidence of a levelling out.

We have seen that stabilizing selection removes the extremes, which include the individuals that are homozygous at most loci affecting the characters. Directional selection, on the other hand, involves breeding from those individuals closest to an extreme, and the elimination of those that manifest the phenotypes less desirable in terms of performance. The maize kernels with the highest oil content will come from plants that contain more genes for that feature. Breeding only from

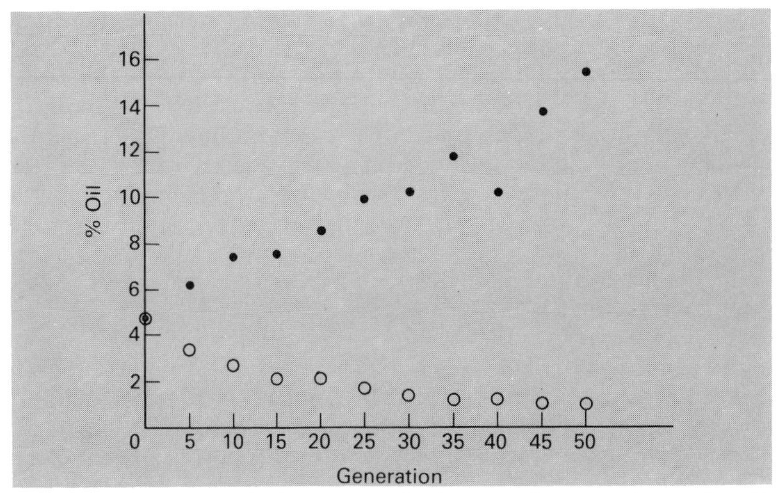

Fig. 5.4 The results of a long-term selection experiment to change the oil-content of maize kernels. From Woodworth *et al.*[272]

these individuals will improve the performance of the line. As selection proceeds, the line will become completely homozygous and, when this situation is reached, further selection is futile. Such a situation is illustrated by a selection programme reported by Robertson[230] and

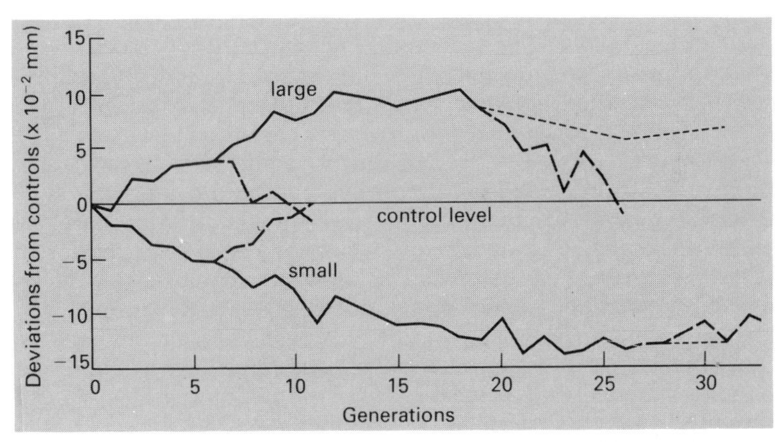

Fig. 5.5 The results of thirty generations of selection to modify thorax length in *Drosophila*. The vertical scale represents the thorax length in hundredths of a millimetre relative to a starting size of zero. The broken lines show the effects upon thorax length of reversing selection, and the dotted lines show the consequences of relaxing selection completely. From Robertson.[230]

illustrated in Fig. 5.5. This experiment involved two lines originating from each of three populations of *Drosophila subobscura*. One line was selected for increased thorax length and the other for reduced thorax. It can be seen that both lines reached apparent plateaux after about 15 generations in all populations. After 5 generations, Robertson attempted to reverse the selection, and succeeded. After 17 generations, he selected in the reverse direction for the high line, again with success. However, when he reversed selection on the low line there was no sign of a return towards the original situation.

It appears that, despite the appearance of a plateau at about generation 15, there is still heterozygosity in the high line at some loci controlling thorax length which can be utilized in reverse selection. It is also apparent that under a situation of relaxed selection the thorax length decreases slightly, presumably under a system of 'natural' selection for an environmentally optimal phenotype. In the low line, however, there is no evidence of a return under either reversed or relaxed selection, so the loci must all be homozygous, and a response to selection has to await mutational change at the loci involved.

This situation often arises in agriculture. A programme of selection to improve the performance of a breed ultimately loses its efficacy, because all the genetic variation is used up. New variation can be produced either by irradiation, or treatment with chemical mutagens, and subsequent renewed selection. Alternatively, agriculturalists can search for wild populations of the species, which have not been subject to experimentation, and may contain alleles that are not present in the domestic stock. One must also remember that a line which has been selected to fixation need not necessarily be homozygous for all the desired genes. Loci are linked together on chromosomes, and an allele at one locus may improve the performance while an allele at a closely linked locus in the same individual may not be the most desirable. Selection may be too rapid to allow recombination to produce the optimum combination, and so the stock becomes homozygous for an allele which is not the best available. Alternatively, desirable alleles may be lost by inbreeding or accidents of sampling during early generations with the same net results. Consequently, crosses between highly selected lines may produce combinations of alleles not present in either parental stock. Genetic reorganization during meiosis, and careful selection can then give rise to individuals which are superior in performance to either parental type. These individuals can be further selected to form new pure-breeding lines. Additional experiments reported by Robertson[230] illustrate this. The populations of *Drosophila melanogaster* in his experiments came from Ischia, an island off Italy, and Crianlarich and Renfrew in Scotland. After reaching the plateau, each of these stocks was crossed with the others, and selected once again for increased size. There was an initial decrease of thorax length, followed by a rapid response to the selection. After a few generations, the thorax size reached a plateau at a level considerably in excess of either parental strain (see Fig. 5.6). These results can easily be interpreted in terms of novel combinations of alleles as we have outlined above. Alleles are fixed in, say, the Ischia line that are absent from the Crianlarich. Crossing them gives a new burst of variation upon which selection can act.

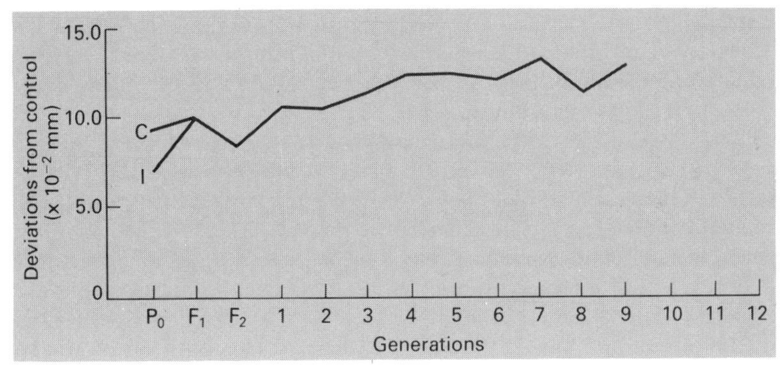

Fig. 5.6 The results of 12 generations of selection to increase the thorax length of *Drosophila*. The starting population is the F_1 between two lines previously selected for increased size until the response had slowed. Population C is from Crianlarich, population I from Ischia. From Robertson.[230]

In conclusion, directional selection is an aspect of quantitative genetics that is of particular interest to agriculture. It is especially amenable to experimental analysis which has shown that it tends to produce phenotypic and genetic uniformity. A knowledge of the mechanisms of inbreeding and selection allows the plant or animal breeder to continue improving his stock by their judicial manipulation. Crossing apparently worked-out lines may result in the release of additional variation that allows new levels of yield to be reached that are far beyond those otherwise available.

DISRUPTIVE SELECTION

The third important mode of selection upon a quantitative trait is disruptive selection. This occurs when a population is simultaneously selected for two or more different phenotypic optima. Individuals which fall into an intermediate category have a lower fitness, or even fail to survive. The most extreme situation is that shown in Fig. 5.2, where the two tails of the distribution survive, and the intermediates do not. This situation ought to promote genetic diversity for in every generation the most extreme individuals survive. These are likely to be homozygous for different alleles, and so heterozygotes are produced.

An experiment reported by Powell[219] demonstrates with remarkable clarity that this may be so. He took a sample of *Drosophila willistoni* from Brazil and used it to initiate a series of replicate population cages. The conditions in the cages differed. Some contained food cups with one kind of nutrient, others contained an alternative, and the rest contained an equal number of each. The food cups were seeded with brewer's yeast or baker's yeast. Some cages contained food cups seeded with both.

Four of Powell's cages had constant conditions of medium and yeast, two cages had both media present, and two had both yeasts. After 15 generations, he assayed 50 individuals from each population to determine the genotype at each of 22 enzyme loci. The proportion of these 22 loci which were heterozygous in each individual is shown in Table 5.2. The average number of alleles per locus is also shown for each population. Both of these parameters reflect the genetic diversity of the population, and both show that there is a significantly greater diversity in the populations which occupy a heterogeneous habitat.

Table 5.2 Genetic variability in populations of *Drosophila willistoni* that were maintained in laboratory cages with different nutrient medium and yeast. From Powell.[219]

Cage	Yeast	Nutrient	Temp.	Heterozygosity	Alleles per locus
1	Bakers	Type 1	25°C	7·29±0·83	1·68
2	Bakers	Type 1	25°C	7·80±0·82	1·68
3	Brewers	Type 1	25°C	8·04±0·81	1·68
6	Bakers	Type 2	25°C	7·96±0·85	1·68
4	Both	Type 1	25°C	11·09±0·89	1·91
5	Both	Type 1	25°C	9·86±0·84	2·00
7	Bakers	Type 1 & 2	25°C	10·16±0·88	2·05
8	Bakers	Type 1 & 2	25°C	9·33±0·88	1·91

These populations are being selected for their ability to survive on two different habitats, and are therefore subject to disruptive selection. The less variable populations are living in relatively uniform conditions.

Disruptive selection has been studied for many years by Thoday and his colleagues at Cambridge University. In one of their major experiments, they established four lines of *Drosophila melanogaster*, two being maintained with flies having the highest numbers of sternopleural bristles, the others with the lowest. The females were always chosen from the 'wrong' line: thus, the lowest female progeny from a high line were always mated with the lowest males from a low line, and the highest females from a low line were mated with the highest males from a high line. In this way there was 50% migration between lines in each generation. Despite this, the populations diverged rapidly (Fig. 5.7). Eventually, there was virtually no overlap between the high and low lines. Skilful manipulation of the material, combined with a sound knowledge of *Drosophila* technology, enabled Thoday and his colleagues to locate several of the polygenes involved in sternopleural bristle number, and to identify their effects. They were thus able to show clearly that these polygenes re-assorted and recombined into supergenes for high and low combinations (refs in Thoday[253]).

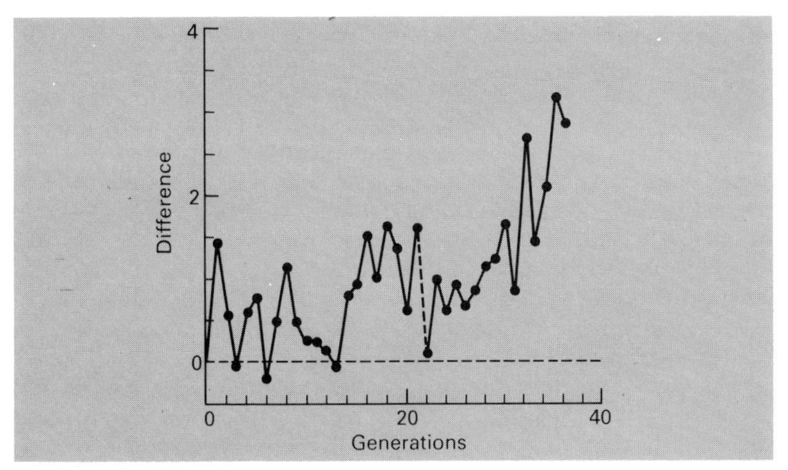

Fig. 5.7 The divergence in mean bristle number between the lines selected disruptively in the population reported by Thoday and Boam (1959). *Heredity*, **13**. There was 50% gene flow between the lines, yet a difference of over 3 bristles arose in less than 40 generations.

However, Thoday and his colleagues produced a more spectacular and important result from another disruptive selection experiment. The situation described above is quite artificial. Enforced crossing of flies from separate lines is genetically interesting, but biologically more difficult to interpret. They therefore allowed random mating among a mixture of equal numbers of the highest and lowest flies of both sexes in a single disruptively selected population. The males were then discarded and the females separated back into the highest and lowest, and placed in vials to lay their eggs. The highest males and females from the high line and the lowest males and females from the low line were then chosen from the offspring and allowed to mate at random once again.

Such an experiment is much closer to a wild situation. If the extreme individuals survive and the intermediate perish, the survivors must come from positive assortative matings. One might expect a behavioural mechanism to evolve so that dis-assortative mating ceased, for adults mating in this way will leave no offspring for the next generation. Only flies which mate assortatively pass their genes on to the next generation.

Thoday and Gibson[254] claim to have observed this. They subjected a population of *Drosophila melanogaster* to this mating and selection regime for 12 generations. They then crossed flies of high and low bristle number to show that intermediates *could* be produced, and that there was no evidence of dominance. They then allowed 8 individuals of each sex from their high and low lines to mate at random, and the offspring which emerged were bimodal for bristle number. The low females produced low offspring, and the high produced high. Thus, it seems that a population subjected to disruptive selection had evolved a behavioural isolating mechanism which ensured the maintenance of two classes of individuals in the population. However, several workers in other laboratories have attempted to repeat this experiment with a conspicuous lack of success.[253] In a thoughtful review, Scharloo[233] suggests that the population used by Thoday to found the experimental lines was genetically heterogeneous, although the situation is still a matter for dispute.[253]

The surprising aspect of Thoday's experiments is not the result, but the speed with which this isolating mechanism arose. To change a mating preference, in some cases after only 7 generations, is a remarkable achievement, and is presumably due in part to the intensity of the selection imposed. There is, in fact, a situation where similarly intense disruptive selection has produced an isolating mechanism in nature. It has been described by McNeilly and Antonovics,[185] but has a rightful place in the next chapter. Consequently, we will defer its discussion until there.

6

Adaptation

INTRODUCTION

In this book, we are attempting to develop some relationships between contemporary studies of population and ecological genetics, and the process of evolution by the accumulation of individually small genetic differences into the gene pools of populations. This accumulation should result in the diversification of the populations, and the evolution of a new species is complete only when these diversifications are sufficiently great as to create a reproductive barrier which cannot be crossed. We have seen that populations can change as the result of natural selection or other agencies acting on one locus at a time, and also how selection can act upon quantitative characters both in the laboratory and in nature. The present chapter will describe some studies which bridge the gap between single loci and quantitative characters, by examining firstly research into systems where there are a few, clearly identifiable, loci. We will discuss the way in which these loci interact to produce more complex phenotypes, and the results of the action of natural selection upon them. We will then proceed to a consideration of genetic systems which cannot be examined locus by locus, and characters which do not fall into neat discrete units. It will become apparent that all of these stdies are linked by the action of natural selection moulding the gene pool to ensure the production of individuals which are in general optimally adapted to the environment in which they will live. The genetic composition and consequently the phenotypic structure of adjacent populations may differ, and in many cases careful study and research has accounted for the differences.

The starting point for a review such as this must be the research into mimicry, which Lewontin[175] called one of the 'great successes' in the study of the causal explanations of genetic polymorphism in terms of differential survival and rates of reproduction of genotypes.

MIMICRY

We have seen how Hubby and Lewontin[127, 176] showed that there may be a large amount of polymorphism in natural populations. They also showed that if loci were assumed to act independently, heterozygous advantage alone could not be maintaining this polymorphism, because the populations would not be able to produce sufficient offspring to overcome the genetic mortality required. The possibility arises that loci do not act independently and we have a certain amount of evidence that this possibility is correct. Some of the best evidence that there are certain combinations of genes which are much more beneficial than others comes from experimental studies relating to mimicry. There have been many pieces of research into this interesting phenomenon in recent years, and the results of a few are distilled to give this brief review.

Batesian mimicry

There are, of course, two kinds of mimicry. The first is named after the great explorer-naturalist, H. W. Bates, whose accounts[12] of the fauna of the Amazon basin in the middle of the nineteenth century make fascinating reading even today. Batesian mimicry refers to the situation where a palatable and harmless species gains protection by virtue of its resemblance to some distasteful or otherwise dangerous species. There have been several experiments verifying this, many of which have been the work of Lincoln and Jane Brower in America. Their earlier studies are typified by one involving captive Florida scrub jays (*Cyanocitta caerulescens*) feeding upon monarch (*Danaus plexippus*) and viceroy (*Limenitis archippus*) butterflies.[19] Four birds were separately offered a choice between a monarch and a butterfly known to be palatable. They had never met monarchs before, but presumably found them distasteful as they rapidly learned not to eat them. Indeed, after an initial learning period, the monarchs were never even touched, although the palatable butterfly was always eaten. This last observation confirmed that the birds were hungry, but that they appreciated that the monarchs were unpalatable.

Four other birds were offered a viceroy in addition to the palatable form. These birds had no experience of monarchs and ate the viceroy often enough for it to be evidently palatable.

When the birds which had been trained with monarchs were offered a viceroy in addition to the palatable control, the latter alone was eaten. The viceroy was never eaten, and in many experiments, it was not even pecked at. This shows that the birds' experience with monarchs was sufficient to protect viceroys from predation. Consequently, we can be sure that the viceroy is an effective Batesian mimic of the monarch. Experiments of this kind indicate that, within the confines of a laboratory cage, mimetic forms will gain a benefit from their resemblance to a distasteful model, provided that the predator has experienced its unpalatable nature.

Attempts to verify this advantage in the wild are more complex to undertake, and less clear-cut in their results. There are two kinds of experiments: the first uses real animals as prey species, the second involves artificial situations with pastry or card prey, with quinine hydrochloride being added to make some distasteful. We will discuss the wild situations involving real animals first as this is biologically more realistic, but the results are rather messy.

The chief experiments are those of Brower, Cook and their colleagues[22, 57] undertaken in Trinidad. Their experimental situation involved painting a palatable species so that it resembled a distasteful model. These animals were then released, and their survival compared with control individuals which were also painted but which did not resemble the model.

The species that they used as a model was the butterfly *Parides anchises*. They showed that this species was distasteful by offering it to several species of birds which all refused it. Their 'mimetic' species was the day-flying moth *Hyclophora prometha* which was palatable to the same species of birds. Female *P. anchises* are generally black with a grey spot on both aspects of the fore-wing and a red spot on the top and a pink band on the bottom of the hind-wing. Male *H. prometha* are also blackish-brown, but the borders of their wings are white, and they lack the grey, red and pink colouration. Brower and his colleagues blackened the borders of these wings with paint, and painted grey, pink and red in the appropriate places. The controls were also painted black at the borders of their wings, but the coloured patches were painted black in some controls and blue in others. The former of these had more or less uniformly black wings, the latter possessed blue spots and bands, which is not a colour generally associated with any distasteful species of butterfly in Trinidad. In the event, there was no difference in the survival of these two kinds of control.

Brower and his colleagues[22] released 414 of their synthetic mimics and 832 controls in one experiment. The moths were liberated over a period of 24 days in an isolated valley in Trinidad. Traps containing virgin female *H. prometha* were put out in the forest, and, apart from differential predation, there is no reason why 'mimics' and controls should not have been recaptured equally.

When they examined their results, Brower *et al.* found that 95 mimics and 254 controls had returned. This means that the controls survived better than the mimics!! Out of such disaster, phoenix-like, scientific advance may arise. So it was with this experiment, for the authors noticed that the frequency of mimics in the recaptured moths was initially high, but declined over the course of the experiment. They suggested that the mimic gained an initial advantage, but that as more animals were released the advantage declined with the changing relative densities of model and mimic. When a mimic is locally common, predators will encounter it sufficiently frequently to change their hunting strategy: black butterflies with coloured spots will, more often than not, be palatable.

Brower *et al.*[22] now modified their experiment by releasing painted butterflies in six different areas and recapturing only for one day. This would not give a density of palatable mimics which was locally too high nor would there be time for the predators to learn that the synthetic pattern was palatable. The experiment was continued at 26 new sites in 1964, 1965 and 1966, and a further 32 in 1967.[57] In the first three years there was a clear and statistically significant advantage to the mimics (Table 6.1). However, in 1967 this advantage disappeared (Table 6.2).

Table 6.1 The number of *Hyalophora prometha* recaptured one day after their release in sites where they had never previously occurred. Those painted to mimic the distasteful *Parides anchises* were recaptured significantly more than the controls. Data from Cook *et al.*[57]

	Recaptured	Released	% Recaptured
Mimics	93	315	29·5%
Controls	72	319	22·6%
		$= \chi^2_{(1)} = 4.0$	

Table 6.2 The results of a similar study to that in Table 6.1, but obtained in a year when the general butterfly community was very much reduced in number. Data from Cook *et al.*[57]

	Recaptured	Released	% Recaptured
Mimics	138	434	31·8%
Controls	157	431	36·4%
		$= \chi^2_{(1)} = 2{\cdot}1$	

In this last year it was generally agreed that the Trinidad butterfly fauna was very much reduced, and the authors suggest that their results in that year stem from this. When there is plenty of alternative food supply, mimics gain a considerable advantage. When the food supply is limited, however, predators become more discriminating and pick out the differences in behaviour between model and mimic.

The results of these studies indicate the complexities of biological experimentation under field conditions. It is possible to obtain consistent results from one series of experiments to the next, but vagaries of the environment can frequently blur the picture so that determining the true facts becomes difficult. For this reason, several workers have turned to more artificial situations to uncover the secrets of Batesian mimicry. There are three principal studies which relate in turn to the relative numbers of models and mimics, their degree of similarity, and the degree of unpalatability of the model. We will discuss them in that order.

The earliest experiments are those of Brower[20] involving the use of mealworms with a coloured band painted around them. The models were painted green and dipped in quinine hydrochloride to render them unpalatable. Mimics were painted identically, but were dipped in water instead of quinine, as were edible or control worms which were painted with an orange band. Captive starlings (*Sturnus vulgaris*) were alternatively offered orange mealworms to test whether they were hungry, and a model or mimic randomly but in pre-determined overall relative frequencies. Birds which received 40%, 70% or 90% models rejected 80% of the mimics without touching them. This indicates that they associated the colour pattern with a distasteful experience. Birds which were given 10% models only rejected 17% of the mimetic mealworms, which suggests that the selective advantages of a mimetic form is frequency-dependent. When the mimic is commoner than the model, it gains less advantage than when it is rare.

O'Donald and Pilecki[207] used small pieces of pastry rather than mealworms and dyed them to achieve differences in colour. They used yellow as the palatable control, and blue and green models which they also made distasteful with quinine hydrochloride. Birds coming to a feeding table rapidly learned to avoid blue or green baits during an initial training period, showing an equal dislike of both. They were then simultaneously presented with five differently coloured baits. In addition to the two models in equal frequencies, there were a rare blue mimic, a commoner green mimic, and a yellow control. Birds pecked at and ate significantly more green baits than blue, suggesting that a rare mimic gains a real advantage over a commoner one.

When the experiment was repeated using a higher concentration of quinine hydrochloride, the results were different. The rarer mimic was now no longer at an advantage over the commoner one. The authors suggest that this shows the selective advantage to be gained depends upon the degree of distastefulness of the model. If it is quite abhorrent, the relative abundance of the mimics is unimportant. They also propose that under a mildly distasteful régime, a mimetic polymorphism might evolve that was frequency-dependent. Rare mimics would be at an advantage over commoner ones which would result in an adjustment of their relative frequencies. The equilibrium situations would depend upon the relative abundances and palatabilities of the models.

Morrell and Turner[198] used a slightly different experimental design in their analysis of the effects of varying the resemblance of the mimic. They made small triangles of card and painted them. These were laid upon the ground with a piece of pastry upon each, apparently looking somewhat like a butterfly. Green cards held a normal piece of pastry as a control, while red cards held a distasteful quinine bait. The birds soon learned to avoid the latter. When birds were presented with controls, models and mimics in the ratio 1:2:1, the mimics were almost completely protected. This we might expect from the earlier studies.

The birds were next presented with imperfect mimics: either red cards with a black bar across, or yellow cards. Both of these were less protected than the perfect mimic, though the red and black was protected more than the yellow. This suggests that colour is more important to the birds, but, significantly, the level of predation of both imperfect mimics increased with time. Just as Brower and Cook[22, 57] found in Trinidad, birds were learning that the mimetic pattern was not precise, and were using this knowledge to distinguish palatable from potentially distasteful prey.

We can now review these experimental results. Batesian mimics undoubtedly gain a benefit from their resemblance to a distasteful model. This benefit declines when the frequency of the mimic increases, or when the degree of resemblance is reduced. Avian predators can certainly learn to make quite fine discriminations between even slightly different prey, and this happens more rapidly when alternative food is in short supply. In a polymorphic situation, a species can mimic two models, and the equilibrium situation will depend upon the relative frequency (and presumably, distastefulness) of the models. The fitness of the mimics themselves will be frequency-dependent, for the benefit will be greater when a phenotype declines below the equilibrium position.

We can now turn to the most intensively studied Batesian mimics in the world, the swallow-tailed butterflies of the genus *Papilio*. Some of these species have been subjected to detailed genetic and distributional analysis by Clarke and Sheppard of Liverpool University. The results are detailed and complex. We will choose a single species and discuss the implications of its variations upon the populations found in one small area of the Orient.

Papilio memnon is a widely distributed swallow-tail from south-east Asia. The male is constant in appearance over the entire range of the species and is non-mimetic. The females are highly variable from place to place, and several polymorphic forms are usually found in any one locality. These sympatric mimetic females usually differ from one another (and the non-mimetic) in a series of characteristics, such as the presence or absence of tails to the wings, the colour of the body, the colour and pattern of fore- or hind-wings, etc. The genetic analyses of Clarke, Sheppard and Thornton[54] have shown that many of these characters are controlled by single loci at each of which several alleles may segregate in a particular population. Furthermore, five of the loci are linked together on the same chromosome. These loci control the presence of tails, the colour of the body, the fore-wing pattern, the hind-wing pattern, and the colour of the 'epaulettes' (which are small coloured spots resembling the coloured 'bleeding points' from which distasteful fluid oozes in many unpalatable butterflies). The consequence of this close linkage is that two butterflies which are phenotypically very distinct may differ from one another as the result of a single 'supergene' of five closely linked alleles.

A clear example of this comes from the fauna of the region around Hong Kong. Here, there are two very similar distasteful butterflies called *Parides coon* and *Pachlioptera aristolochiae*. These two species resemble one another in the colour and pattern of the fore-wing, and both have a yellow abdomen, red epaulettes and tails to the hind-wing. There are three female forms of *Papilio memnon* in this area. One (*distantanius*) mimics *P. coon* and another (*alcanor*) mimics *P. aristolochiae*. The third form (*agenor*) is non-mimetic; it lacks tails, has a black abdomen and a differently shaped white hind-wing patch.

The genetics of these three forms has been disentangled by Clarke *et al.*[54] who show that there are three chromosomes present in the population of *P. memnon* from Hong Kong. These carry genes for:

A—tailed	:	*distantanius*	hind-wing	:	yellow body
B—tailed	:	*alcanor*	hind-wing	:	yellow body
C—tail-less	:	*agenor*	hind-wing	:	black body

These can combine to give 6 genotypes whose phenotypes are as listed below:

AA — mimics *Parides coon*
AB — probably mimics *Parides coon*
AC — mimics *Parides coon*
BB — mimics *Pachlioptera aristolochiae*
BC — mimics *Pachlioptera aristolochiae*
CC — non-mimetic

Crossing over within the supergene in AC and BC parents will produce novel chromosomal alignments, and recombinant offspring which in fact have been obtained in the laboratory. They are very rare in the wild, however. It is perhaps significant that they are more often produced from pupae collected in the wild than from adult insects captured on the wing, for these recombinants will be imperfect mimics and we have seen that avian predators can readily detect minor imperfections in the mimetic phenotype. There is a complication to this argument, however, for there is a naturally occurring non-mimetic form, and this occurs at quite a high frequency in nature. Thus, Clark *et al*. report that of a sample of 43 *P. memnon* from Hong Kong, 33 were of the non-mimetic *agenor* type. We have suggested that non-mimetic recombinants are at a selective disadvantage, and yet *agenor* clearly is not. The possibility of long-term heterozygous advantage of the supergene 'alleles' can be ruled out for recombination can occur, and presumably would break up these balanced complexes. It is not known what other factor might maintain the *agenor* form in the population, but its very existence argues that a selective advantage must accrue to it at some stage in the life-cycle, and that recombinant individuals are selectively removed by bird predators.

This mimetic situation is similar in *P. memnon* populations from other parts of its range, and hinges upon the close linkage of the loci involved. It is a moot point whether the variations arose as mutations on separate chromosomes, and then became situated upon a single one by translocation. It seems likely, however, that closer and yet closer linkage would be favoured by selection to reduce the wastage of gametes in the production of selectively disadvantageous non-mimetic recombinants. The evolution of a supergene with particular alleles included within it will be a matter of considerable selective importance. Furthermore, the dominance relations of the alleles involved are total. In the example from Hong Kong, all the *agenor* characters are recessive. Were some dominant and others not, the mimetic resemblance would again collapse, for AC and BC would be neither *agenor* nor mimetic. Again, the ability of predators to discern minor variations from the distasteful model would result in excessive predation of the less mimetic form.

Mullerian mimicry

In the previous section we discussed the way in which a species can gain protection by virtue of its resemblance to a distasteful model. There is another form of mimicry, however, which has been named after F. Muller who first described it in 1879.[199] This is the situation where a series of distasteful, or otherwise protected, species gain a mutual advantage by resembling one another. A simple example would be the yellow and black striping on the abdomens of many bees, wasps and hornets. If a predator attacks one of these and receives some punishment, all members of the Mullerian mimetic complex will gain a slight benefit. It is possible, of course, for palatable species to be Batesian mimics of a Mullerian complex: many species of harmless Diptera possess the black and yellow stripes of a wasp. But true Mullerian mimics are all unpleasant in one way or another.

There have been few experimental studies of Mullerian mimicry, although one of the best concerns a series of palatability trials of butterflies from the sub-family Heliconinae reported by Brower *et al.*[21] These animals are widespread in tropical and subtropical parts of the new world, and there are several pairs of species which resemble one another.

The experimental procedure adopted by Brower and her colleagues was fairly straightforward. Silverbeak tanagers (*Ramphocelus carbo*) were captured and kept in cages. Twenty heliconid butterflies of a particular species and twenty of another species that was known to be palatable were offered one at a time, but in a random order. Acceptance or rejection was noted, and it was found that the birds rapidly learned to reject the heliconids—suggesting that they were distasteful.

Brower *et al.* found that some of the birds rejected heliconids right from the start of the experiment, which implied that they already knew that such butterflies were distasteful. In fact, offering a 'generalization' heliconid butterfly to a bird after it had experienced a particular species showed that an advantage accrued based upon size and shape alone. Not surprisingly, the advantage was greater if the 'generalization' butterfly was more similar. Brower *et al.* argued that if silverbeak tanagers had already experienced heliconid butterflies in the area of Trinidad where they were captured, rejection ought to be most strong for the species which are most common in the same area. Less abundant species are less likely to have been experienced and so should be less advantaged. This indeed was so. Fifty per cent of the locally abundant species of heliconid were rejected by birds on the first occasion that they were offered, as opposed to only 22% of the less common species.

It seems from these results that Mullerian mimicry works in the experimental situation. Birds which are offered a distasteful species will reject similar distasteful species when they are encountered. This phenomenon will result in selection to improve the resemblance between distasteful species. Fisher[93, 96] argued tht if there were two similar and distasteful species, the less common would gain a greater benefit from its resemblance to the more common, but that this form would also gain a slight advantage every time the less common was tested and rejected by a predator. This is a crucial point. In Batesian mimicry, the advantage of the warning colouration of the model is reduced whenever a mimic is killed and eaten. But in Mullerian mimicry, all members of the complex are advantaged every time one of their members is tested by a predator.

Selective predation will also promote uniformity among Mullerian mimics. Any individuals that stand out from the norm will be recognized as distinct by avian predators and attacked. Sometimes, they will be killed despite their lack of palatability. Selection of this type will remove those individuals farthest from the optimum phenotype and uniformity will be increased. Polymorphism should not occur, for segregating varieties may be sufficiently distinct to be subjected to increased predation.

In an interesting study, Turner[257] discussed the variation of two distasteful butterflies. *Heliconius melpomene* and *H. erato*. These species coexist over much of tropical South America, and Brower *et al.*[21] showed that they were both distasteful. Turner lists ten forms of each species which differ in the colour and pattern of their wings. In general, only one form of each species is found in any area, and the variety distribution of the two species is almost identical. It seems that in any one area, predators are exerting selection to improve the resemblance between the two species to form a remarkable Mullerian complex. No form of one species can spread into the area occupied by another form of the same species without destroying the advantage accruing from the mimicry. The two species are tied to one another apparently ineradicably, and Turner suggests that the pattern of distribution which he describes stems from a few jungle refuges during the last Ice Age. Within each refuge, the action of avian predators upon these two species produced particular combinations of alleles giving rise to virtually identical phenotypes. Following the amelioration of the climate, the vegetation changed and allowed the two species to spread until the various populations came into contact. Hybridization occurs along the boundaries, but the phenotypes produced are no longer Mullerian mimics and so are removed by predators.

SUPERGENES

Mimicry is a supreme example of the interaction of natural selection and polymorphic loci, but it also gives us great insight into a most interesting phenomenon that we have touched upon before. The five loci that control the basic phenotype of *Papilio memnon* in the Hong Kong region of Asia are linked together so closely that recombination only occurs very rarely. Such a group of alleles is effectively inherited as a single unit, and is often called a supergene. This term is now in general use to describe the situation where a group of loci are closely linked together and behave to all practical purposes as a single locus. We have come across a supergene earlier in *Cepaea nemoralis*, for, in this species, the ground colour of the shell and the presence or absence of bands upon it are similarly controlled by two very tightly linked loci.[30]

The evolutionary advantage of such a system of supergenes is readily appreciated by considering the consequences of the independent assortment of the loci involved. In the *Papilio memnon* populations around Hong Kong, there are three common phenotypes that are controlled by three distinct supergenes. A mimetic individual that carries the non-mimetic recessive supergene is heterozygous at three loci controlling the phenotype. Were these loci not linked, eight different gametes could be produced, only two of which would be present in the parental generation. Over half of the offspring would be imperfect mimics, and consequently be selectively disadvantageous. The presence of the loci in a tightly knit unit removes this evolutionary undesirable situation.

This is an example of a supergene being maintained by the selective action of predators, but they can arise in other ways. A classic example of supergenes is to be found in the heterostyle polymorphism among plants. Probably the most intensively studied example relates to the primrose (*Primula vulgaris*) of Europe. This species commonly occurs in two forms called pin-eyed and thrum-eyed which differ principally in the structure of their corolla. These are shown in Fig. 6.1, and differ most obviously in the position of the anthers, and the stigma. Pin-eyed plants have a long style, so that the stigma is situated at the opening of the corolla tube. The anthers are sited midway between the stigma and the nectaries. In thrum-eyed plants, these organs are reversed: the anthers are at the top of the corolla tube, and the style is short.

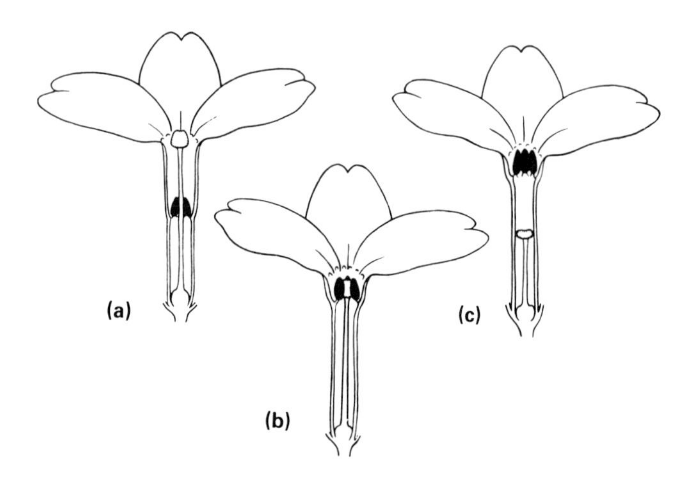

Fig. 6.1 Three varieties of the primrose, *Primula vulgaris.* The top left individual is a pin flower having a long style and low positioned anthers. The top right is a thrum with a shorter style, and the anthers situated at the top of the corolla tube. The middle form is a tall homostyle, with anthers at the top, and a long style. From Ford.[102]

Charles Darwin undertook a pioneering study of this system, and pointed out that it promoted cross-breeding. *P. vulgaris* is frequently visited by butterflies and bumble-bees, both of which have a sufficiently long proboscis to reach the nectaries at the base of the corolla tube. Pollen adheres to the proboscis freely, but especially the parts which are adjacent to the anthers. Thus, insects visiting pin-eyed plants collect pollen midway down the proboscis, whereas visitors to thrum-eyed plants get most of the pollen at the base of the proboscis. Subsequent visits to the same form of plant result in a build-up of pollen at these points, and relatively little is transferred to the stigma. When a plant of the opposite type is visited, a large mass of pollen is now adjacent to the stigma, and effective pollen transfer takes place.

There are other morphological differences between the two types of *P. vulgaris*. The surface of the stigma of pin plants is rounded and covered with long papillae. Thrum plants have flatter stigmas and very short papillae. The pollen grains differ in size, being appreciably larger on thrum-eyed plants. This last result has been used by Ford[102] to confirm the differences in the position of pollen transport by pollinating insects. He states that the larger grains from thrum-eyed plants are predominantly carried close to the base of the proboscis, whereas the smaller pin pollen is lower down towards the tip.

The two forms also differ in physiological attributes. The purely mechanical effects of differences in the anatomy do not, of course, prevent like mating with like: they merely reduce the probability. It appears, however, that there are real fertility barriers within the fertilization process after pollen has been deposited upon the stigma. Thrum pollen will germinate upon a thrum stigma, but is normally unable to penetrate the surface. Pin pollen can both germinate upon and enter a pin stigma, but its growth down the style is very slow. Sufficiently slow to be always beaten by a thrum grain growing alongside. The papillate nature of the stigmas may play a part in the penetration of pollen grain, and the differences in size of the grains themselves may reflect variation in the amount of reserves available for pollen tube growth.

The inheritance of the pin and thrum phenotypes initially appears straightforward. The thrum form is dominant to the pin form. Because of the virtual absence of self-pollination SS is almost never produced and only two genotypes are present

(Thrum) Ss and ss (Pin)

This situation is not unlike the sex-determining systems of so many organisms. There are two genotypes which can only mate with each other, and thus replicate their own genotypes in equal proportions.

More detailed analysis by Ernst[90] revealed that there are a series of closely linked loci rather than one major gene. These loci control anther-height, style-length, pollen-size, the length of the papillae, and the penetration and rate of pollen tube growth. Linkage is extremely close, and again there is an apparent supergene. Recombination can, and does, occur in nature, for 'homostyle' primroses have been found established in two populations by Crosby,[59] although they also occur at a very low frequency in many other places. Recombination appears to be capable of producing two homostyle forms, one with long style and high anthers (which is the form found by Crosby) and the other with short style and low anthers.

Both homostyles ought to spread widely through those populations into which they are introduced. The anthers and stigmas lie in close proximity and the sequence of the loci in the supergene is such that a recombination produces the pin male characters in association with thrum female, or *vice versa*. That they have not spread widely is a matter of concern to the *Primula* ecological geneticists, and is still under dispute (for a review see Ford[102]).

LINKAGE DISEQUILIBRIUM

We have seen that the colour and banding loci of *Cepaea nemoralis* are closely linked. Although there is still some uncertainty, it seems likely[202] that the recombination rate is less than 1%. The alleles at these loci are usually inherited as supergenes, and it is possible to make a simple estimate of the frequency of these, in addition to the major genes themselves.

For example, Cain and Sheppard[32] report a sample which they collected from a population in Tackley Heath Wood near Oxford. This sample consisted of 60 yellow banded snails, 111 pink bandeds and 79 pink unbandeds. There were no browns, and the yellow unbanded phenotype was also absent. Table 6.3 lists the frequency of each morph, and also the supergene composition of the phenotypes. If we assume that the Hardy-Weinberg law holds in this population (and there is no evidence either way on this score), we can write down the frequency of every 'super-genotype' and accumulate them into their respective phenotypes. These can be equated directly with the phenotype frequencies derived from the field data, and solved to give only one set of supergene frequencies that are biologically realistic.

$$s = \text{YB} = 0.490$$
$$r = \text{YO} = 0$$
$$q = \text{PB} = 0.337$$
$$p = \text{PO} = 0.173$$

If the population from which this sample was taken had been in equilibrium, the proportion of banded to unbanded individuals would have been the same in both colour classes. Furthermore, the chromosomes themselves can be classed as 'pink' or 'yellow' depending which colour gene they carry. The proportion of pink chromosomes which carry the banded gene would be the same as the proportion of yellow chromosomes. We can estimate the expected proportion of each chromosome (effectively each supergene) directly by the product of the constituent gene frequencies, as follows:

$$\text{Expected frequency of PO} = 0.510 \times 0.173 = 0.088\ (0.173)$$
$$\text{Expected frequency of PB} = 0.510 \times 0.827 = 0.422\ (0.337)$$
$$\text{Expected frequency of YO} = 0.490 \times 0.173 = 0.085\ (0)$$
$$\text{Expected frequency of YB} = 0.490 \times 0.827 = 0.405\ (0.490)$$

It is apparent that these depart markedly from the actual frequencies in the population, which are printed in brackets. The situation where a combination of linked alleles are present in a population at a frequency significantly above or below expectation is called linkage

Table 6.3 The estimation of supergene frequencies in a sample of *Cepaea nemoralis* reported by Cain and Sheppard.[32]

	Number	Frequency	Constituent Genotypes
Yellow banded	60	0·240	$\frac{YB}{YB}$
Yellow unbanded	0	0	$\frac{YO}{YO}$ $\frac{YO}{YB}$
Pink banded	111	0·444	$\frac{PB}{PB}$ $\frac{PB}{YB}$
Pink unbanded	79	0·316	$\frac{PO}{PO}$ $\frac{PO}{PB}$ $\frac{PO}{YO}$ $\frac{PO}{YB}$ $\frac{PB}{YO}$

Let the frequency of the four supergenes be as follows:

$$PO = p \quad PB = q \quad YO = r \quad YB = s$$

Then from the Hardy-Weinberg law, the phenotypic frequencies will be:

Yellow banded	s^2	$= 0{\cdot}240$
Yellow unbanded	$r^2 + 2rs$	$= 0$
Pink banded	$q^2 + 2qs$	$= 0{\cdot}444$
Pink unbanded	$p^2 + 2pq + 2pr + 2ps + 2qr$	$= 0{\cdot}316$

Whence, the most biological realistic estimates of p, q, r and s are:

$$s = 0{\cdot}490 \quad r = 0 \quad q = 0{\cdot}337 \quad p = 0{\cdot}173$$

disequilibrium, or sometimes gametic phase disequilibrium, because the gametes are being produced in non-equilibrium proportions. It can arise from several different causes, some of which we will now discuss.

In the sample of *Cepaea* that we have been discussing, there is clearly a substantial deficiency in the yellow unbanded supergene. This can be accounted for by the effects of visual selection by predators acting to promote crypsis. Tackley Heath is a deciduous woodland which has a carpet of dead leaf litter and herbs at different times of the year. The background is predominantly dark or broken, and the uniform yellow appearance of a YO shell will be very conspicuous. This form will be removed by predators to a disproportionate extent. Because recombination is so low, new YO supergenes will be produced very slowly from YB/PO parents, while the deficiency will increase every time a YO snail is killed and eaten.

In this way, selection can produce and maintain a linkage disequilibrium, in the face of recombination producing the 'missing' genotype. This is a general rule, and is not restricted to the supergenes of *C. nemoralis*. It is the reason why there is presently so much interest in genetic disequilibria in population genetics.[37,189]

Selection is not the only mechanism for establishing a linkage disequilibrium, however. It can be produced by a method somewhat akin to genetic drift. It can sometimes happen that a population is drastically reduced in size by some catastrophic, but non-selective, mortality. The survivors might vary slightly from the equilibrium situation—indeed as with genetic drift it is probable that they will. When the population increases in size again, the differences from equilibrium will be enhanced, and closely linked loci may take a long time to regain equilibrium. Theoretical studies reviewed by Crow and Kimura[61] cast light upon the approach to quasi-linkage equilibrium, so-called because the nature of the algebra renders the approach asymptotic, and theoretical equilibrium is never reached. They thus work in terms of 'half-life' or the time taken to go halfway to equilibrium. This time depends closely upon the recombination rate. From total disequilibrium, the half-life is seven generations with a recombination rate of 10%, but over 690 generations if the recombination is 0.1%.

These results make the assessment of a linkage disequilibrium situation difficult in nature. Only if we know that bottlenecks have not occurred in the population for many hundreds of generations can we be sure that the disequilibrium is selectively based. In the case of *Cepaea nemoralis* around Oxford, we know very little about the past history of the populations. However, without considering visual predation, we can deduce that selection is important because there is a deficiency of unbanded yellows, compared with pink, in over 80% of Cain and Sheppard's woodland samples. This could reflect a single massive bottleneck in the last Ice Age which has resulted in disequilibrium in every population descended from this sole ancestral unit. However, the non-woodland samples from the same area show deficiency in yellow unbanded in only about 40% of samples. These populations are mixed among the woodland ones fairly randomly. Such a consistent difference between the two habitats makes the single bottleneck argument unlikely.

This kind of methodology only has to be used when it is not possible to establish a functional relationship between the phenotype and the selective forces. The disequilibria manifest in the populations of *Papilio memnon* around Hong Kong discussed earlier in the chapter can be explained selectively. Certain supergenes are at a distinct disadvantage because they do not produce a mimetic butterfly. Similarly, recombinant forms of the heterostyle supergene in *Primula* may be at a long-term disadvantage because self-fertilization is possible, and inbreeding depression sets in.

Finally, it is possible for linkage disequilibria to be produced quite artefactually by sampling a subdivided population. It sometimes happens that a population is structured, possibly being divided by some slight barrier to gene flow. When this happens differences in gene-frequency can arise within the area normally occupied by a single randomly-mating population. The two sub-units (or demes) may be in linkage equilibrium within themselves, but differ in the frequency of the constituent supergenes. Thus if one has a high frequency of Ab, and the other of ab, an apparent deficiency of Ab and aB will be produced in the sample quite spuriously. So, in addition to understanding the past history of the populations and the existence of bottlenecks in the past, one must know a great deal about the population ecology of the species involved. Here, truly, ecology and population biology are complexed together with the theories of population and evolutionary genetics.

LINKAGE DISEQUILIBRIUM IN EXPERIMENTAL POPULATIONS

While discussing the problems of closely linked loci and their effects upon population genetics, it is perhaps relevant to consider a major difficulty which they pose to the experimental analysis of natural selection. When a species is established in the laboratory, the parents used as foundation stock are frequently a small sample of the entire natural population. Consequently, linkage disequilibria are likely to be established which may last for many generations and interfere seriously with experimental manipulation of the artificial population.

Yarborough and Kojima[282] report upon a series of experiments which were designed to test whether natural selection operates upon the esterase-6 locus in *Drosophila melanogaster*. This enzyme has two common allelic forms called 'fast' (F) and 'slow' (S) from their mobility upon an electrophoretic gel. Yarborough and Kojima established their base population by crossing two inbred lines, and maintained it for 30 generations in a large cage. By this time, the fast allele had attained a frequency of about 30%. In generation 31, they isolated 14 lines that were homozygous for the S allele and 6 for the F allele, from single pair matings. These lines were increased in number by transfer to replicated cages and used in all subsequent experiments.

They then established 8 experimental cages using all 20 lines. Four of the cages commenced with high frequencies of fast allele, and four with low. They were then left for a further 30 generations, after which 7 populations showed a movement in the frequency of fast towards 30%, which was the equilibrium point of the base population, (see Fig.

6.2). They claimed that the 20 lines would differ at most of the loci apart from esterase-6, and that the reversion towards the equilibrium point was due to selection acting upon this locus alone.

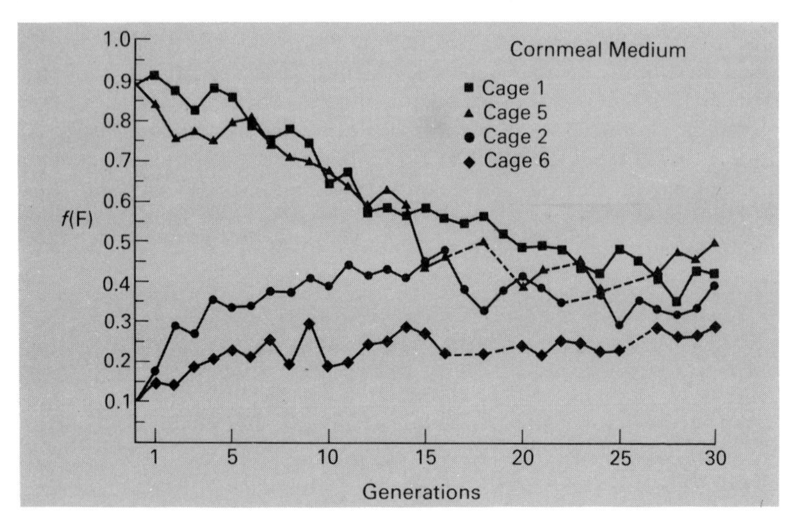

Fig. 6.2 The changes in allele frequency at the esterase-6 locus in four experimental populations of *Drosophila melanogaster.* Two replicates were started with a high frequency of the F allele, and two with a low frequency. Note the convergence in these four populations towards a broadly similar frequency. From Yarbrough and Kojima.[282]

From the theory outlined by Crow and Kimura,[61] any loci within $2\frac{1}{2}$ recombination units of the esterase-6 locus will only have moved half-way towards their linkage equilibrium states after 30 generations. Thus, the base population which they so carefully established will, in fact, still be in substantial disequilibrium. The 20 lines will similarly be affected, and there is a real possibility that the results which they obtained were due to closely linked loci, and not to esterase-6 at all.

Jones and Yamazaki[137] report upon a similar series of population cages of *D. pseudoobscura*. They monitored the frequency of the fast and slow alleles at the sex-linked esterase-5 locus. Cages established with only two founding lines showed marked temporal changes in frequency, whereas those started from 10 or 20 lines showed far less. These results were in general agreement with an earlier study by Yamazaki[281] who found no fluctuation at all if 44 lines were used to initiate the population cage. Jones and Yamazaki suggest that they have demonstrated the importance of the rest of the genome (the genetic background) in influencing fluctuations in gene-frequency at enzyme loci.

In a paper delivered to the 13th International Congress of Genetics, Clarke[46] has stressed the importance of applying the 'mechanistic link' methodology to the study of enzyme loci. He argued that only by linking the biochemical attributes of the enzymic variants to population genetics can selection be proved. Circumstantial evidence can then be provided in support of the basic proof, but circumstantial evidence alone is not initially sufficient.

Clarke then proceeds to apply this system to the alcohol dehydrogenase locus (Adh) in *Drosophila melanogaster*. There are two alleles at this locus in most populations, and these again are called fast (F) and slow (S) from their electrophoretic mobility. It is generally agreed that the two alternative homozygous forms differ markedly in their biochemical properties. Rasmusen *et al.*,[222] Gibson[108] and Day *et al.*[70] all showed that the enzyme produced by the fast homozygote has a greater specific activity than the slow when ethanol is used as the substrate. Day *et al.* showed that this held for a variety of other alcohols as well.

It seems likely that alcohol dehydrogenase acts to detoxify alcohols, in which case the differences in activity could reflect differences in the rate of detoxification. This should be a factor of considerable selective importance, and Morgan[197] attempted to test the prediction that survival might differ between genotypes on different alcohols in a manner related to the rates of detoxification. He established two population cages, one homozygous for FF and the other for SS, using 20 independently derived lines each. He then set up a series of competition experiments with 200 of each homozygote on a food medium which included one of a series of alcohols examined by Day *et al.*[70] He found that the survivorship of FF relative to SS was greater on those alcohols which had a higher FF/SS activity ratio (see Table 6.4). This result was in agreement with the prediction.

It is also relevant that Gibson[108] found that the fast allele increased in frequency in populations which were kept on ethanol-enriched medium. Finally, 1-pentene-3-ol is an alcohol which is converted to a highly toxic ketone. If this alcohol is added to the food medium in a similar competitive situation, there is a higher emergence of SS flies.[197] The argument here, which could be (and was) predicted, is that the faster activity of the FF flies gave them a disadvantage because there was a more rapid accumulation of toxic ketone in their bodies.

There is one more piece of information which concerns not the activity but the stability of the enzymes. At high biological tempera-

Table 6.4 The relationship between enzyme action and genotype survival in *Drosophila melanogaster*. The 'relative activity' column reports the activity of the Adh enzymes produced by FF homozygotes relative to the SS enzymes in a series of alcohols. The remaining three columns relate some results of competition experiments between the two genotypes in media containing each alcohol. From Morgan.[197]

	Relative activity of enzymes FF/SS	Number of flies surviving out of 400 of each		Survival of FF flies relative to SS
		FF	SS	
Cyclohexanol	2·1	120	12	10·0
Ethanol	1·8	83	17	4·9
N-butanol	1·5	80	22	3·6
Iso-butanol	1·3	88	31	2·8
N-propanol	1·2	98	57	1·7
Iso-propanol	2·5	130	91	1.4
Control	—	60	49	1·2

ture, the slow migrating enzyme is more stable. It was shown by Vigue and Johnson[261] that the frequency of the slow allele increases from north to south in the U.S.A. Furthermore, Pipkin *et al.*[215] showed that the frequency of the F allele increases with altitude in Mexico. Alcohol is present in the food of many *D. melanogaster*, for they typically feed and breed in fallen fruit, which decays slowly. In warmer localities, either towards the southern limits of the species, or at lower altitudes, the more thermolabile fast enzyme might lose its activity, in which case the FF flies will be less able to deal with the toxic alcohols being taken in with the food. These flies will die, and the frequency of the slow allele will correspondingly increase. Evidence in support of this argument comes from a study by Johnson and Powell[132] who showed that heat shock favours the slow allele, and cold shock the fast in experimental situations.

The evidence is now becoming appreciable that natural selection can act upon the alcohol dehydrogenase locus, moulding and changing the gene frequencies to an optimum for the environment in which the population lives. Time will tell whether other enzyme loci behave similarly, or whether alcohol dehydrogenase is an exceptional locus within the genome.

THE EVIDENCE FOR ADAPTATION IN METRIC CHARACTERS

The advantage of qualitative variation to studies of genetic variation and evolutionary change are considerable. The methods of inheritance are usually fairly straightforward, and consequently the phenotype can often be related directly to the genotype. Change in frequency of alleles can be observed and quantified, which makes the disentanglement of the controling and peripheral variables potentially straightforward. We have seen examples of this many times already, and have also discussed how more complex interacting loci can similarly be examined and sometimes explained in terms of the interplay of selective forces. The situation becomes more difficult when one turns to metric or quantitative characters. There is a vast amount of this continuous variation which is not easily resolved into a few alleles and their frequencies. Nevertheless, as we saw in Chapter 5, characters such as size or flowering time are subject to a high degree of genetic control, and as such are potentially selectable. The rules of selection and random genetic drift must apply to the alleles controling metric traits and so it ought to be possible to detect consistencies in the distribution of phenotypes which, at least in some cases, will be relatable to the external environment.

A suitable starting point was chosen by Bradshaw and his colleagues during the 1950s. They determined to examine morphological variation in plants which inhabited markedly different physical environments. Their aim was to establish whether there were genetic differences between the populations, and to uncover the selective or other forces responsible. To be able to relate the genetics to the environment in a mechanistic way is the goal of studies in this field, and Bradshaw and his co-workers have succeeded spectacularly.

They chose the grass *Agrostis stolonifera* for their early work. This species is a perennial which is almost exclusively wind pollinated. It grows profusely on the coastal grasslands of North Wales, and Aston and Bradshaw[7] examined populations that grew around Abraham's Bosom. They found (Fig. 6.3) that plants growing close to the sea were short and compact, whereas those from more sheltered sites were tall and slender.

To eliminate the possibility that these differences were due to environmental modification of the phenotype, Aston and Bradshaw took tillers, or cuttings, from the field plants and grew them under controlled conditions, in a greenhouse. The differences in morphology between the populations were maintained, which implies that they were truly genetic in origin. They also collected seed from plants in the wild and raised this in standard greenhouse conditions. Again, the seed resembled the population from which it came, although there was greater variability among the seed plants than the tillers, especially if they came from sheltered conditions.

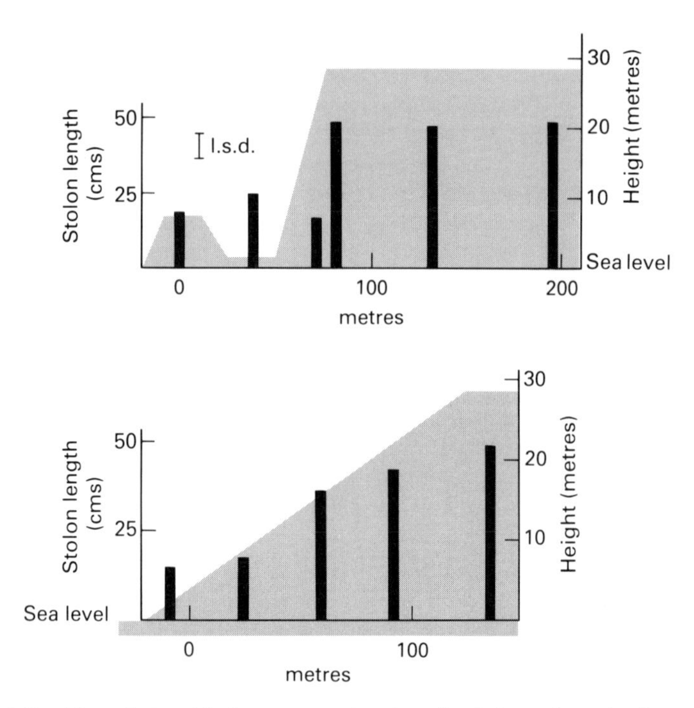

Fig. 6.3 The relationship between stolon length of *Agrostis stolonifera* and position on transects up two hillsides through Abraham's Bosom. Stolons were measured as the means of tillers taken in the wild and reared under standardised greenhouse conditions. Note the sharp discontinuity in the dimensions of plants from the 'sudden change' transect. From Aston and Bradshaw.[7]

Observation and measurements in the field indicated that the wind was much stronger on the exposed hillsides close to the sea than in the more sheltered gullies and ditches close by. Aston and Bradshaw showed experimentally that the long straggly *A. stolonifera* plants from sheltered sites were badly damaged when exposed to the wind. The shorter, tufty plants from sea cliff situations were virtually unharmed by this treatment, and flowered and set seed well even after a whole winter of exposure to the elements.

These results, both ecological and experimental, led Aston and Bradshaw to suggest that selection by wind had favoured the survival and reproduction of short plants in exposed situations. Over the course of time, the genetic composition of the populations that inhabit such sites has been modified accordingly. The genes producing tallness or slender stature have been reduced in frequency by the failure of such phenotypes to reproduce. Conversely, in sheltered sites, the plants need to be taller in order that they might compete successfully with other herbs and grasses for sunlight, and also to remain above water should the ditch habitats become flooded. Once again, natural selection has modified the gene pool of the ditch and gully plants to produce the optimum phenotype for the environment.

The difference in variance shown by seed and tiller plants can also be accounted for on this hypothesis. The adult plants from which tillers were taken had gone through the process of selection and survived. They represented the optimum phenotypes for that locality at that time. The seeds, however, represented an unselected sample of plants for that environment. Prevailing winds in Abraham's Bosom come from the sea, and so some pollen is carried from exposed sites into the sheltered gullies. Seed produced from this pollen will be less tall and so these hybrids, which would probably fail to survive in the wild, grow and mature in the greenhouse to increase the phenotypic variance of the sample.

It might be argued that such extreme differences between habitats are the exception rather than the rule. More typically, the environment will change gradually from one condition to another and the selective pressures will correspondingly be reduced. One might conclude that major differences between environmental parameters are perhaps as atypical as the major genes of large effect which manifest themselves in visible differences in the phenotype of snails and moths. Consequently, they are equally likely to be involved in strong selection.

Recently, however, Watt[269] has provided a clear example of how animal populations can become genetically adapted to a gradually changing climatic variable: in this case temperature. As a rule, there are more severe fluctuations in daily temperature within a single locality than there are between the means of situations several hundred miles apart.[107] Nevertheless, there are trends towards decreasing mean temperature both with increasing latitude and altitude, and Watt showed that butterfly populations showed clear evidence of their adaptation to this variable.

There is a genus of butterflies named *Colias* that has several species and sub-species in the western parts of North America. They vary both within and between species in the colour of their wings, particularly in the relative amounts of yellow and black (see Fig. 6.4). Watt saw that populations from high altitudes or northern latitudes were generally darker in coloration than the others. He reported[269] that behavioural and physiological studies showed there to be no

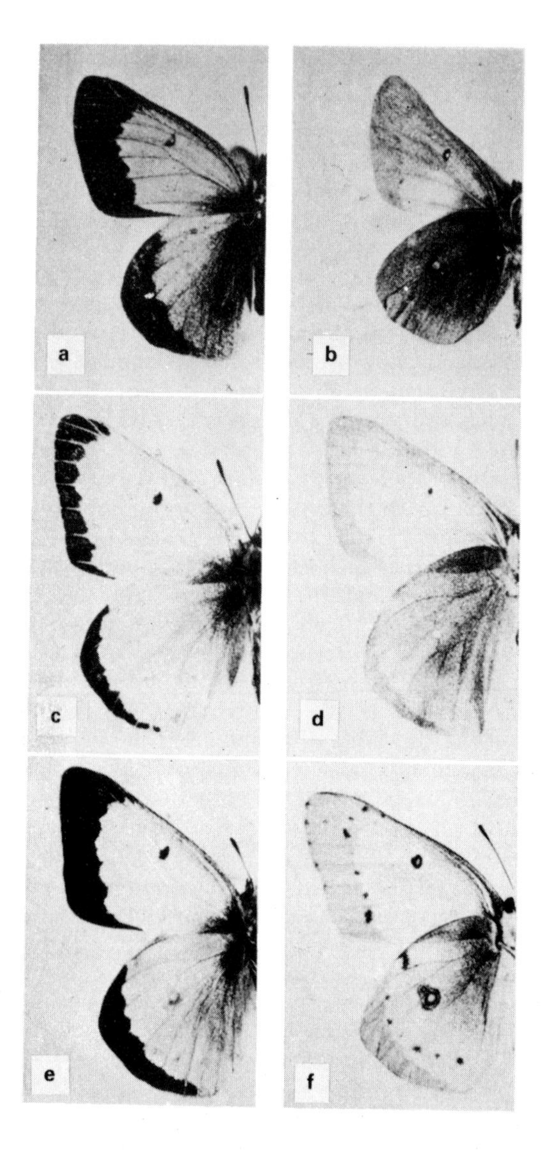

Fig. 6.4 Male *Colias* butterflies from an altitudinally zoned series from British Columbia. The upper drawings show the upper side, and the lower show the underside. **a** and **b**, *Colias nastes*, alpine; **c** and **d**, *Colias palaeno*, montane; **e** and **f**, *Colias alexandra christina*, lowland. Notice that the high altitude forms are distinctly darker. From Watt.[269]

internal mechanisms for temperature regulation. Apparently, they do not use muscular activity to raise their internal temperatures nor evaporation to lower it. By implanting thermistors in the thorax of living butterflies, Watt was able to show that active flight usually took place when the thoracic temperature is in the range of 28–42°C. Below this level, the animal is too cold to be able to take flight, and at higher temperatures heat distress becomes apparent. He found no evidence that the more northerly or higher altitude butterflies were adapted to fly at lower temperatures, either external or thoracic. Watt also reported differences in the behaviour of *Colias* depending upon the thoracic temperature. When this was low, they orientated in a plane at right angles to the incident thermal radiation, but at higher temperatures, they stood parallel to the radiation. In this way they were able to maximize or minimize the heat being absorbed by the body.

He then matched pairs of butterflies for sex, size and weight, and found that dark ones warmed up faster than light coloured ones. This happened whether they were alive or not, suggesting it to be a physical rather than a biological phenomenon. The dark butterflies also stabilized at a higher equilibrium temperature, whatever the environmental conditions (see Fig. 6.5).

It is apparent from Fig. 6.4 that the dark parts of a *Colias* wing are

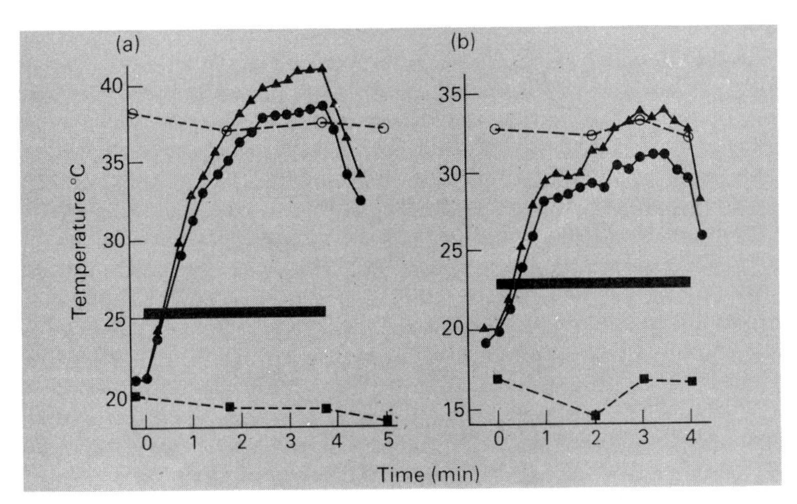

Fig. 6.5 The results of two experiments showing heating rates of size and sex matched pairs, one light coloured and one dark, of *Colias* in perpendicular orientation to the sun's rays. (**a**) *Colias scudderi* (dark) compared with *C. philodoce* (light). Live insects. (**b**) Dark and light *C. eurytheme.* Insects dead 14 days at time of experiment. From Watt.[269]

those closest to the thorax, and Watt suggested that the relative amounts of black and yellow were adaptive responses to the environment. In the cold northern or high altitude situations, darker animals are at a selective advantage, because they heat up faster in the morning and can take flight earlier. They remain more active because they maintain a higher temperature during the day, and so they can feed and, more importantly, mate to a greater extent than the paler individuals. The genes which confer dark pigment will consequently increase in frequency within the populations. Conversely, in warmer situations, dark butterflies overheat, and so have to spend the hottest part of the day in heat avoidance postures. At moderate temperatures, this involves orientation parallel to the sun's rays. But at more extreme temperatures, they are compelled to seek out shady places where they can hide away from the sun. In either case, the time available for feeding and reproduction is reduced, and so the frequency of the pigment-forming genes will decrease. The population will thus become more yellow.

Over the course of time, this ebb and flow of advantage and disadvantage will mould the gene-frequencies at the pigmentation loci towards the optimum for the locality. Immigration from adjacent populations will tend to reduce the level of adaptation, but stabilizing selection will subsequently eliminate those individuals that depart markedly from the most suitable phenotype. The gene pool will gradually evolve towards a situation which might be termed equilibrium—where gross departures from the optimum phenotype are minimal. Whether or not Watt's populations have attained this state is not clear, time alone will tell. Indeed, we do not really know how long a period is necessary for responses to selection to be apparent, although two recent studies suggest that it might in some circumstances be quite short.

The first of these concerns further research by Bradshaw and his colleagues into the effects of strong selection upon plant populations. In addition to his work on wind exposure, he has also led a programme of study into the effects of heavy metal contamination of the soil upon the ecological genetics of grasses. This work has been primarily based upon the plant species that live on the spoil heaps of heavy metal mines. The waste material from these mines usually contain too little extractable metal to be economically valuable, but sufficient to provide a highly toxic substrate for plants. Consequently, little grows on the piles of rock and rubble around copper, tin, zinc or lead workings. The few plants that do grow there are extremely interesting, for they are tolerant to much higher concentrations of heavy metals than members of the same species from uncontaminated sites.

Bradshaw[15] discussed a population of lead tolerant *Agrostis tenuis* growing on Goginan mine in North Wales. He reported that plants taken from nearby pasture would not grow on mine soil. Root growth was almost totally suppressed, yet the plants originating from Goginan mine itself showed the ability to overcome the toxicity of the soil.

Tolerance was measured more precisely by Jowett[138] who grew both seed and tillers in a water culture containing a solution of the appropriate heavy metal. Tolerance was recorded as the relationship between root growth in toxic and normal solutions. It became apparent from these studies that many of the lead mines of Wales, both in use and abandoned, supported populations of *A. tenuis* that were tolerant to lead, and furthermore that the tolerance was heritable. Gregory and Bradshaw[115] showed that plants which were tolerant to lead were only tolerant to other metals if they were also present in the soil from which the plants come. And, as a general rule, the tolerance mechanism seems to be metal specific.

Further light was cast upon the ecological genetics of heavy metal tolerance by a study of Drws-y-Coed copper mine by McNeilly.[184] He took samples of seed and tillers from *A. tenuis* growing in a series of stations on two transects across the boundary of this mine. He also recorded the level of copper in the soil. It is difficult to define 'toxic' soil, because the quantity of copper present in a sample does not necessarily correspond to the amount available to the plant. Copper can be bound to large organic molecules in the soil, which renders it inaccessible and so reduces the soil toxicity. However, on the basis of copper concentration and the presence of plants in the sampling area, the toxicity of the soil was estimated at each sampling position, and

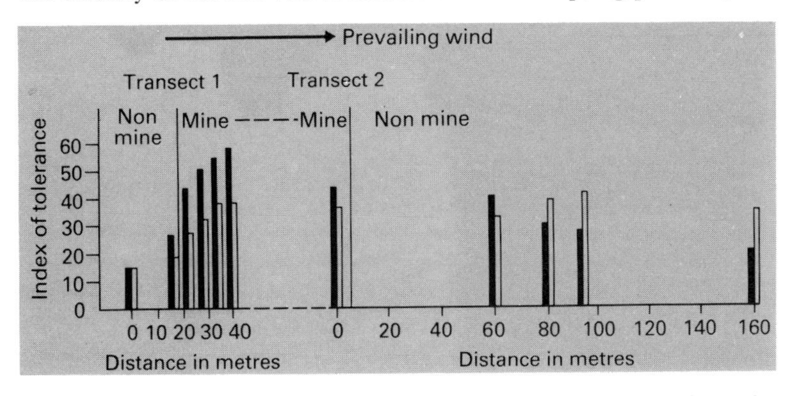

Fig. 6.6 Mean index of copper tolerance in *A. tenuis* of a series of samples taken in two transects across Drws-y-Coed Mine. ■ adult population sample, □ seed population sample. From McNeilly.[184]

those samples taken from inside the mine boundaries were found to come from highly toxic soil. Figure 6.6 shows the spatial relationships of the samples and their tolerance estimates.

In transect 1, there is a rapid change in the tolerance of the adult plants that corresponds fairly closely to the estimate of copper toxicity. This is not unreasonable. Only tolerant plants can survive on the mine conditions. However, the pattern of seed tolerance across the boundary is rather different. There is an increase in the mean, but it is considerably less than that in the adults. McNeilly accounts for this by recourse to gene flow. He notes that the mine lies in a valley with a strong directionality to the wind. Transect 1 is at the predominantly up-wind side of the mine. Consequently, a large amount of pollen is carried onto the mine from the non-tolerant pasture plants outside. This pollen has the effect of reducing the tolerance of any fertilized seed, and so the mean declines. These plants with a lowered tolerance would almost always fail to survive on the mine soil, and so the adult population has a higher mean level of tolerance. This is shown well in Fig. 6.7, which illustrates the index of tolerance of the tillers and seed taken in the samples at site 4 on transect 1. This sample was taken very close to the mine boundary, but on toxic soil. The overwhelming majority of seeds will be fertilized by immigrant pollen, and this is reflected by 62.1% of the seed having an estimated index of tolerance less than 30%. Almost all of this seed is doomed, however, for only 13.3% of the adults have tolerance in this range and the difference reflects the consequences of selective mortality.

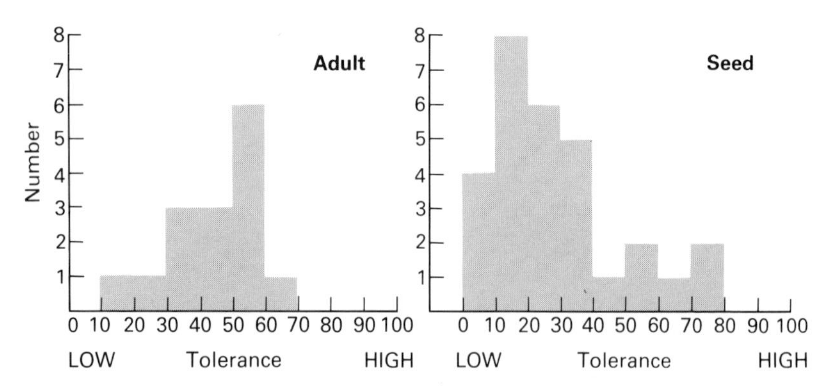

Fig. 6.7 The profile of copper tolerance at site 4 of transect one in Fig. 6.6 (asterisked). Note that seed is less tolerant because some has been fertilised by pollen produced by plants growing off the mine. The adults reflect a post-selection sample, and therefore are more tolerant to environmental copper. From McNeilly.[184]

In transect 2, the pattern is different. The samples are now from the down-wind side of the mine, and so tolerant pollen is being carried onto the pasture plants. This raises the tolerance of plants living off the mine, and the seed is more tolerant than the adult plants. This difference in tolerance can again be accounted for by selective effects. McNeilly showed that, in a mixed population, non-tolerant plants produced more dry weight and more tillers than tolerants when both were grown in normal soil. There is thus an advantage to non-tolerant plants which is reflected in the pattern of tolerance found in the samples of transect 2. Plants growing further away from the mine are still sometimes being fertilized by tolerant pollen, but tolerant seed is less fit and so is less represented in the adult population.

The change at the boundary is more dramatic in transect 1 because non-tolerant plants die on the mine. The selective forces are thus more extreme than in transect 2 where tolerant plants *can* survive on normal soil. The difference in the latter case is that the tolerant plants grow more slowly, and so the selective forces are less strong.

At first sight, it appears that the evolution of this metal tolerance has occurred very rapidly for many of the mines have been present for less than 150 years. However, it is possible that the area was contaminated with heavy metal before mining operations started, and so precise assessment of this point is rather difficult. However, there is another example of the rapid differentiation of a species, and one which can be allocated a more definite time scale. This concerns the evolution of the house sparrow (*Passer domesticus*) following its introduction into North America.

The house sparrow arrived in North America sometime between 1852 and 1860, although the number of individuals on introduction is not known. It found the continent to its liking, and rapidly spread so that within a hundred years it was found from coast to coast, and from the Great Lakes to Mexico. During the period of this spread, there is no change apparent in the ancestral European populations. Johnson and Selander[136] report that samples from Europe taken in 1852 and 1962 do not differ statistically in wing length, tail length or plumage characters. Consequently, any variation in the American populations of house sparrows is probably due to evolution taking place within these populations alone, and not part of any world-wide trend.

Johnson and Selander[136] have analysed and reported at length upon the variation to be found in contemporary populations from North America. They took samples from 31 sites upon the mainland, and also two islands, Bermuda and Oahu, Hawaii. Their initial analysis concerns 16 skeletal characters, in particular the lengths and diameter of various long bones, a series of skull characteristics and the dimensions of several thoracic and pelvic bones. They have not formally proved that these parameters are under genetic control, but it seems extremely likely that they are.

The statistical methodology in this paper is detailed and complex, but the results can be easily summarised. In both sexes they found a trend in almost all skeletal characters reflecting an increase in gross body size with decreasing winter temperature. This trend did not relate to summer temperature at all. They also found a second relationship resulting from an increase in body core and a decrease in long bone size, which correlated with both summer and winter temperature.

Results such as these are well known from zoogeographical studies of many species of warm-blooded animals, and relate to the thermal interactions between individuals and the environment. Many species of homeotherms increase in size with latitude, and the phenomenon is known as Bergmann's rule. The amount of heat produced by an animal is roughly proportional to its mass, and thus to its volume. The rate at which heat is lost to the outside world depends upon its surface area. Consequently a large homeotherm will maintain an above-ambient temperature more efficiently than a smaller one, always assuming that it can find enough food to maintain itself alive. Allen's rule, on the other hand, relates the size of peripheral appendages to the ambient temperature. Extremities tend to get smaller, relative to the body core, in colder climates, because they are major sources of potential heat loss.

North American house sparrows demonstrate both Bergmann's rule and Allen's rule. The morphological variation involved must have evolved during the spread of the species into new areas within the 110 years following its introduction into the continent. Evidently, genetic variation at the loci controlling these morphological traits must have been present in the founding population, for it seems that such a short period is insufficient for the relevant alleles to have arisen by mutation. However, Klitz[159] has now examined serum and tissue protein polymorphism in North American house sparrows. He reports that there was virtually no serum polymorphism in a total of 303 birds from 10 different sites, nor among 122 individuals assayed for tissue protein. He estimates that 20 loci were examined.

This result is in marked contrast to most of the surveys of enzyme and other protein polymorphisms which have been undertaken. However, Selander and Kaufman[234] found no polymorphism in the colonizing land snail *Rumina decollata*, and we have seen that Avise and Selander[8] found very little in cave populations of *Astyanax*. Furthermore, Prakash *et al.*[221] found less polymorphism in the isolated, and probably recently founded, Bogota population of *D. pseudoobscura*. Thus, the house sparrow is not along among colonists in its depauperate genetic complement. Nevertheless, it is even more remarkable that the North American populations parallel so faithfully the morphological variation manifest in many other species of homeotherms.

We do not know the level of polymorphism among European populations of species, but nevertheless it seems likely that the small founding population of house sparrows was indeed homozygous at many of its loci. Perhaps some enzyme alleles were lost by genetic drift in the generations immediately following its introduction, and while the populations were still relatively small. Reduction in variation at metric loci may be less likely following a founding incident because of the increased number of loci involved. The situation is complex and needs further study.

Alternatively, it may be that the morphological variation is influenced by regulating or controlling genes. Almost nothing is known about gene control in higher organisms, but this result suggests that different kinds of genes may be involved in enzymic and metric characters. The result is one of fascinating interest and considerable evolutionary importance. If, for example, the loci controlling gene expression are highly duplicated, novel polymorphism could arise in a homogeneous isolated population either by the increased number of mutations which would occur, or by recombination or other reorganization within the duplicated sections of the chromosomes.

Finally, Selander and Johnson's study has cast light upon one aspect of the vexed problem of the rate of evolution. In his review of evolution in animals, Mayr[183] suggests that sub-speciation occurs more rapidly among mammals than birds, and quotes examples of the divergence of populations following their separation by geographical events such as inundation of land areas, or the separation of oceans by newly arisen land masses. We will return to this in the final chapter, but comment here that the evolutionary changes in North American house sparrows are so great as to almost warrant sub-specific separation at the present time, and yet they have arisen in only 110 years. They made us revise our opinions of how rapidly evolution of complex characters may occur, and yet serve as a reminder of the power of natural selection in some systems.

7

Coadaptation

ADAPTIVE LANDSCAPES

We have seen how populations can become closely adapted to the environment that they inhabit. Natural selection weeds out the substandard forms, and the relationship between phenotype and genotype results in an increase of advantageous alleles and a decrease in the less desirable. Consequently, any offspring produced by an individual or population have a chance of resembling quite closely the phenotypes that were successful in previous generations. In a relatively stable environment this is clearly of advantage to the species, and when migration between populations is limited it will result in the production of a series of locally adapted micro-geographical races. We can see an example of this in a map of phenotype frequencies of *Cepaea nemoralis* in the Oxfordshire region.[32] Here, the populations are fairly well separated in a patchwork of woodlands and open habitats. Movement is restricted in land snails, so many of the populations are genetically isolated from one another, and the pattern of shell colour distribution mirrors quite faithfully the habitat background (see Fig. 7.1).

Among more mobile species, movement between adjacent areas is increased and the discontinuous nature of the variation is blurred. Figure 7.2 shows a map of the frequency of the melanic form of *Biston betularia* in north-western England.[14] There are areas of high frequency of the melanic phenotype in the urban areas of Manchester and Liverpool with more or less steep clines between them. These clines are unlikely to reflect solely the intermediate nature of the site in question; they will also to some extent indicate the immigration of

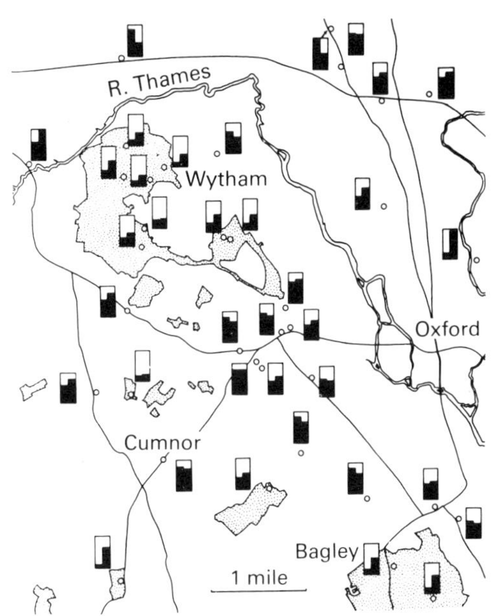

Fig. 7.1 The frequency of yellow shells (left hand column) and effectively banded shells in a series of samples of *Capaea nemoralis* taken from the Oxford district of lowland England. Note that in this sedentary species, the patchwork pattern of habitats is reflected in the discontinuous distribution of phenotype frequencies. From Cain and Sheppard.[32]

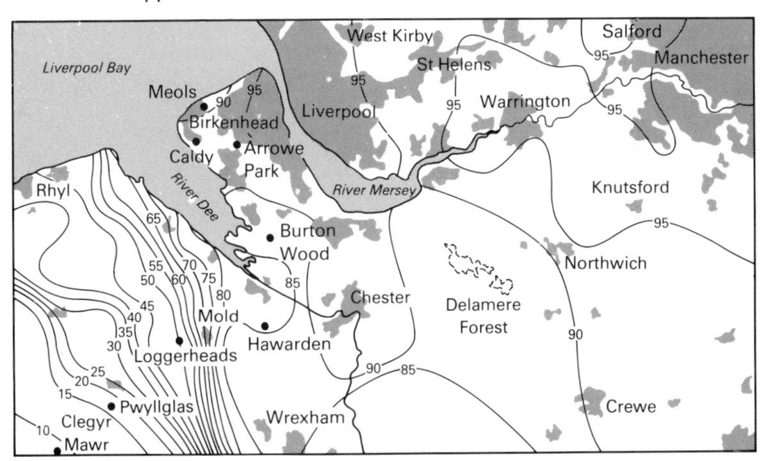

Fig. 7.2 The frequency of melanic *Biston betularia* in north-western England and north-eastern Wales. This relatively mobile species shows a much smoother, more clinal, pattern of phenotype frequency between urban and rural populations. Contrast this with Fig. 7.1. From Bishop and Cook.[14]

melanics from one direction and typicals from the other. Nevertheless, within any area, the population will usually be at, or adapting towards, its own particular local optimum.

The situation of locally adapted genetic entities has been likened to a geographical map, with hills and valleys connected by hillsides and slopes of varying steepness. This concept of 'adaptive topography' was initially developed by Sewall Wright[276] and has been extended by Dobzshansky.[78] The simplest example would have two loci, each possessing a series of alleles. These could be depicted as a two-dimensional array with each combination of alleles being represented by a point. The fitness of these combinations could be then represented as a third dimension giving a model resembling a topographical map. The model can be extended into a more complex state, if an extra dimension is added for each additional locus. However, this rapidly becomes impossible to picture mentally, although the analogy remains useful.

Returning to the simplest case (Fig. 7.3), contour lines can be added to connect combinations of equal fitness, and the resemblance to a topographical map is enhanced. The peaks, or areas of high altitude, indicate where particular gene combinations are advantageous, and valleys or depressions show the converse. Closely aggregated peaks might signify a group of closely related species which have rather similar biologies. More distant summits reflect the position of species which show rather different biological strategies.

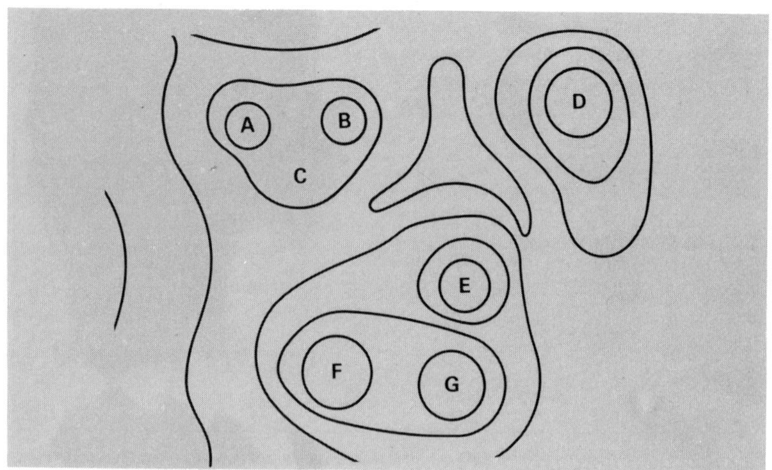

Fig. 7.3 An adaptive topography showing several species on their adaptive summits. Some species are more closely related, and so occupy peaks that are more adjacent than others.

To move from one summit to another involves passing through a valley of poor adaptation and low fitness. Thus, the gap between two selected species, feeding on different food plants, may involve the production of a form which can feed upon neither. Alternatively, it might represent the classic example of a sterile mule as the consequence of a mating between a horse and a donkey. Both of the parent species occupy an adaptive summit, but the hybrid sits firmly in a valley of low fitness. If we extend the model to a populational rather than a specific level, we can envisage populations that are well-adapted to their environment occupying summit positions. Inter-populational hybrids might then possess less advantageous combinations of genes and be situated in adaptive troughs.

COADAPTATION

A population sitting at the top of one of these adaptive summits will simultaneously be adapted to a whole series of external forces. Some of these will involve simple polymorphic systems with clear external influences. Other attributes of the phenotype will be controlled in a more complex way. We have already seen an example of complex interactions between loci in the work of Turner[257] on Mullerian mimicry. *Heliconius melpomene* and *H. erato* resemble one another closely over much of tropical South America. The pattern of their wings varies from one area to another, but it is always similar in the two species, although the underlying genetic basis is frequently different. In the extreme western strip of Colombia, between the Andes mountains and the Pacific Ocean, both species have blue wings with a prominent red flash. The loss of any of the genes producing this phenotype will reduce the resemblance from which the two species gain their mutual benefit. The fitness of an individual depends upon it possessing *all* the genes necessary to produce the blue and red pattern. The fitness does not increase gradually as each additional gene is added, and so it is more akin to a threshold situation: it is certainly not multiplicative.

When a high biological fitness depends upon the interaction between a series of alleles, they are described as 'coadapted' genes. The term seems to have been coined by Dobzhansky during his studies of the fitness properties of various chromosomal inversions of *Drosophila*. As we shall see, he found that the fitness of an inversion genotype (or karyotype) could vary depending upon the origin of the genes from which it was composed.

The principal species that was studied is *D. pseudoobscura*, which has a variety of inversion patterns on the third chromosome. Some of these inversions are widespread and can be found in many populations; however, they have names that usually reflect the area in which they were first found (e.g. Arrowhead, Chiricahua, Treeline). By comparing the position of the bands in the inversions, it is possible to deduce their evolutionary history, and these have been related to one standard pattern.[78]

In one series of experiments using this species, Dobzhansky and Pavlovsky[79] established four replicate populations in cages, using flies that originated from Pinon Flats, California. These populations included two inversion types, 'standard' (ST) at a frequency of 20%, and 'Chiricahua' (CH) at 80%. In all four populations, there was an asymptotic increase in the frequency of the ST inversion towards 80% (see Fig. 7.4). The trajectory of the curves agreed remarkably well with a model based upon constant fitness values and heterokaryotype advantage.

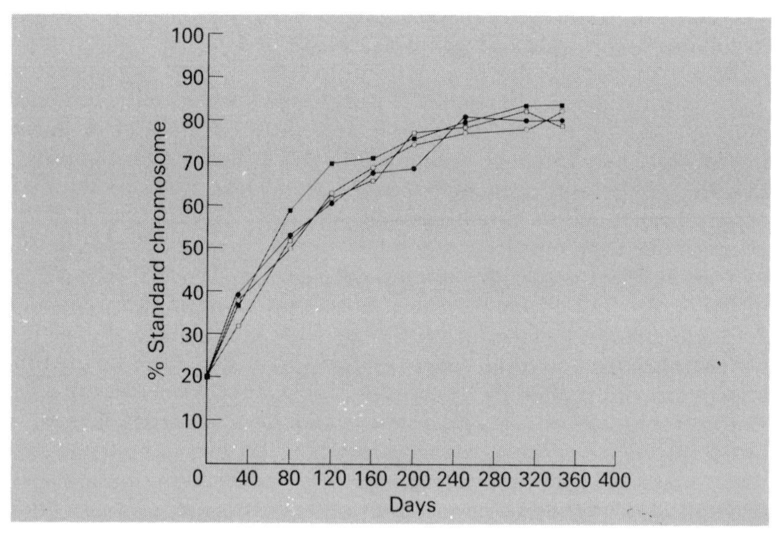

Fig. 7.4 The frequency of the standard chromosome in four replicate populations over 350 days. All four replicates were started with 20% standard and 80% chiricahua from the same population. From Dobzhansky and Pavlovsky.[79]

Wallace[267] lists the results of several other experiments using various pairs of inversions. When the competing pairs come from the same populations, there is only one experiment out of 11 which proceeded to fixation for one inversion. The other ten all attained a stable internal equilibrium. There are seven experiments undertaken by Dobzhansky[75] that are essentially similar with the important exception that the competing chromosomes come from different populations. In all of these, the population moved to fixation of one or other of the inversion types, as shown for example in Fig. 7.5.

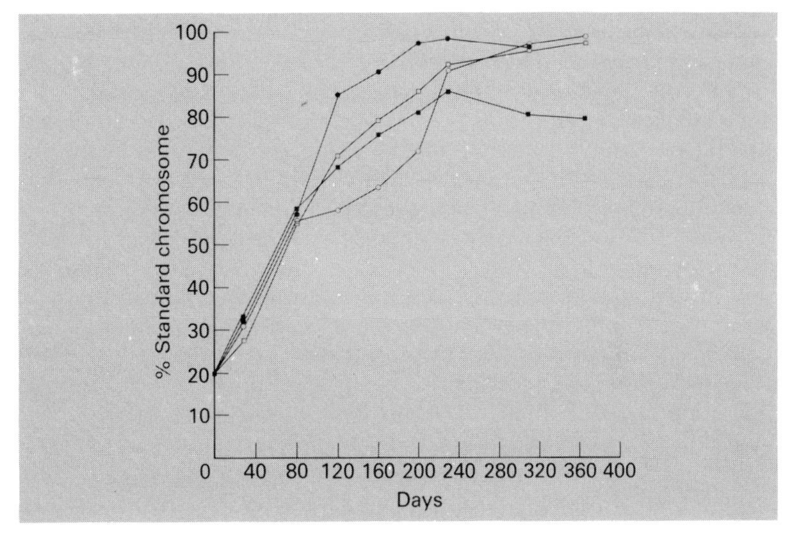

Fig. 7.5 A similar experiment to that shown in Fig. 7.4, except that the standard and chiricahua chromosomes come from different populations. There is no evidence here of the stable internal equilibrium shown in Fig. 7.4. From Dobzhansky.[75]

On the basis of experiments such as these, Dobzhansky[76] suggested that the contents of inversions were coadapted. He argued that one inversion might contain a series of alleles A_1 B_1 C_1 D_1, and the other A_2 B_2 C_2 D_2, which gave a superior heterozygote. Recombination between these groups of alleles would produce alternative combinations that had a reduced fitness, unless combined with the complementary recombinant. For example, A_1 B_1 C_1 D_2 would only have a high fitness when associated with A_2 B_2 C_2 D_1. The containment of these combinations of alleles within an inversion has the effect of preventing recombination, for precise chromosomal pairing is not so straightforward when the order of the genes is different. Dobzhansky considered that the results of these cage experiments supported his views. Stable equilibria usually result when both competing chromosomes come from the same population. When the origins are different, however, the population moves fairly rapidly towards fixation for one or other inversion type. Presumably, if, for example, an ST inversion from one locality has a gene sequence of A_1 B_1 C_1 D_1, that from another may be A_3 B_3 C_3 D_3. There is no reason why these should give rise to a superior heterozygote, and consequently there is no population stability in the experimental cages.

These experiments of Dobzhansky have shown that individual inversions may include coadapted genes, which are held together as a single unit because recombination is prevented. It is possible that such coadapted complexes need not of necessity be associated with inversions. Any population will be subjected to processes such as mutation, inbreeding, random fluctuations in gene-frequency and selection, all of which will alter the genetic composition in their own individual way. Some polymorphic loci or groups of loci, whether linked or not, will be subject to selection, and their gene-frequencies will change accordingly. Others may be selectively neutral, either temporarily or permanently, and the gene-frequencies of these will drift in a random fashion, although if they are linked to selectively important loci their fluctuations may be somewhat constrained. Mutation and migration will occur constantly, producing continual small changes of frequency which may be either maintained or lost.

In this way, following the colonization of a new area, or some environmental change in an existing situation, the whole genetic system of a population will enter a state of flux. Some genes will increase or decrease in frequency, while others are held steady. These may be as polymorphisms under some system of balancing selection, or as monomorphisms because no allele has been produced which does not decrease the fitness of its bearer. There will also be secondary interactions. Genes which increase in frequency may influence other loci, either directly, by virtue of their linkage, or more distantly by affecting the fitness of other genotypes.

Provided that the environment is reasonably stable, this dynamic and evolving state of flux will move towards some sort of 'equilibrium' where selectively important alleles have reached their optimal frequencies, and the less vital ones drift harmlessly up and down. Reorganization of the linkage groups may occur resulting in the creation and spread of new combinations of alleles, but essentially the population will have reached an optimal genetic structure producing individuals of a maximum average fitness for the environment. Lerner[169] discusses some aspects of this process in his brilliant book '*Genetic Homeostasis*' and argues that heterozygosity is to be expected at many of these loci because of their 'buffered' nature, and the consequent reduced departure from optimum fitness.

Under these 'equilibrium' conditions, the population may be

regarded as a single coadapted complex, so that random dips into the gene pool produce individuals of maximum mean fitness. If this is the end product, we can make a concrete prediction about the effects of disturbing such a complex: it should result in a deterioration of fitness, if not in the first generation, then certainly in the second. The first generation may be excluded, for we have seen in the chapter on quantitative characters that a cross between two artificially selected lines may result in an improvement in performance. Advantageous dominant alleles may be present in one population but lacking from the other. Crossing will bring these dominants together which could improve the performance of the hybrid. Furthermore, heterozygosity is likely to be increased in an inter-population cross with a further possibility of hybrid superiority. We need not be surprised if this occurs, but recombination can occur in the hybrids and new combinations of genes will be produced in the gametes, which unite to give zygotes neither heterotic nor coadapted. Here, we might predict a breakdown in fitness.

In principle, it is not difficult to test this prediction, and several workers have undertaken such experiments. Since the next step in the evolutionary process relates to the establishment of isolating mechanisms between populations and their establishment as separate species, these experiments are of crucial importance. Because of their importance, and because of the conflicting nature of some of the results, we will discuss and assess several of the studies that have been made.

The earliest work almost inevitably concerns species of *Drosophila*, partly because of the experimental convenience of these animals, and partly because of the massive technology available to the researchers. It is possible to obtain a series of chromosomes each of which both carries a genetic marker and incorporates a mutation to prevent recombination. Applying these to the experimentally convenient fact that recombination almost never occurs in male *Drosophila* anyway, it is possible to construct individuals containing almost any combination of chromosomes taken straight out of the wild (see Fig. 7.6). This expertise has been widely used in the experiments discussed below.

Brncic

Stocks of *D. pseudoobscura* had been established in the laboratory at Columbia University from single wild-fertilized females that had been captured by Dobzhansky in 1950. During the next three years, a laborious and involved series of crosses was undertaken producing a series of strains of flies, each carrying third chromosomes from one

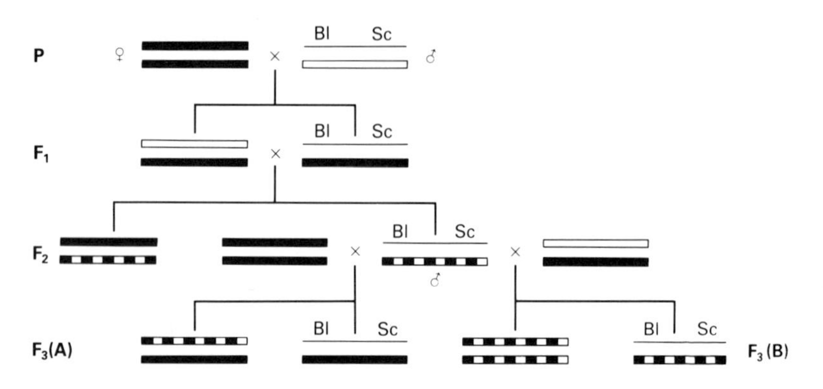

Fig. 7.6 The method used by Brncic[17] to produce flies of known chromosome complement in his analysis of the effects of heterozygosity and recombination on fitness. Chromosomes of different geographical origin are shown as black or white, and recombinants are shown as black and white. The marker chromosome carrying 'blade' and 'scute' is shown as a single line. Note that every progeny segregates for BlSc and wild type. Note also that the BlSc individuals are genetically identical in the F_1 and the F_3(A) crosses, and also in the F_2 and F_3(B) crosses. The wild type fitnesses can thus be compared in these pairs of progeny through the BlSc individuals.

location against a residual genome from another. Simultaneously, strains were produced which carried a third chromosome marked with a dominant gene against the same genetic background. The dominant gene produced clearly distinguishable phenotypes.

These strains were then crossed in a regular way so that there were always two classes of segregating offspring. The first of these was a marked control genotype which had precisely the same genetic composition every time. The second was an unmarked experimental fly carrying a pre-determined pair of third chromosomes against the desired background. The viabilities of these experimental flies could be compared *via* the control offspring, by counting the number of marked and unmarked flies emerging from any cross.

Brncic[17] produced four kinds of experimental flies, which varied in the origin and nature of their third chromosomes. In the first there were two chromosomes from the same location. The second kind of fly carried two chromosomes from different locations. In the third, one chromosome 3 was intact while the other was a recombination product of a female heterozygous for two chromosomes 3 from different locations. The fourth contained two chromosomes 3, both of which were recombination products of such heterozygotes. We will follow Wallace[267] in referring to such chromosomes as 'intact' and 'recombinant'.

Brncic showed that flies whose chromosomes 3 were intact but which came from different locations had a superior viability to flies whose thirds came from the same place (see Table 7.1). Flies with one recombinant chromosome were less viable than either, and flies with two were less viable yet again. The implication of Brncic's results are probably that in the natural state chromosome 3 evolves to carry a series of coadapted genes. When intact chromosomes are put into a single individual, heterotic effects can be observed. But, if recombination is allowed to take place, then the coadapted nature of chromosome 3 is disrupted, and the fitness levels drop rapidly. If both chromosomes are recombinant, then the loss of fitness is greater than with a single recombinant chromosome.

Table 7.1 The frequency of wild type individuals in the progeny classes P, F_1, F_2, $F_{3(A)}$ and $F_{3(B)}$ from Fig. 7.6 in flies having an otherwise identical genetic background. Data from Brncic.[17]

Origin of 'black' chromosome	Origin of 'white' chromosome	P	F_1	F_2	$F_{3(A)}$	$F_{3(B)}$
Mather	Bryce	47·95%	52·08%	50·74%	51·35%	47·37%
Mather	Gunnison		52·19%	44·43%	48·69%	42·64%
Bryce	Mather	48·70%	52·59%	48·81%	49·08%	41·13%
Bryce	Gunnison		54·05%	49·68%	46·33%	41·94%
Gunnison	Mather	49·66%	55·34%	49·40%	46·65%	45·21%
Gunnison	Bryce		51·97%	50·16%	48·98%	44·25%

Vetukhiv

Vetukhiv[258–60] undertook a series of studies whose aims were similar to Brncic, but with simpler experimental designs. Vetukhiv merely took strains of *Drosophila* of different geographical origin and crossed them. He mated male and female *Drosophila* from the same or different populations, and studied their offspring in both the F_1 and F_2 generations. He measured the viability,[258] fecundity[259] and longevity[260] of these flies.

In the first experiment, he placed equal numbers of experimental and genetically marked control larvae in a small volume of medium. By counting the number of wild type and marked adults that emerged, he could estimate the larval to adult viability and compare different experimental flies using the controls as a standard. His results showed (Table 7.2) that in *Drosophila pseudoobscura* and *D. willistoni*, the F_1 hybrids between geographically distinct populations were superior in viability to the parental flies. The F_2 individuals, however, were usually inferior. Similar experiments with *D. paulistorum* were less convincing, for only one of three inter-population crosses showed hybrid superiority.

Table 7.2 The survival rate of various populations and inter-population hybrids of three species of *Drosophila*. In each case, the percentage survival of larvae grown under standard conditions is shown relative to that of a control genotype reared alongside in the same vial. From Vetukhiv.[258]

D. pseudoobscura			
POPULATION	A	B	C
Parental	$\frac{60{\cdot}3}{30{\cdot}0} = 2{\cdot}01$	$\frac{60{\cdot}0}{27{\cdot}9} = 2{\cdot}15$	$\frac{54{\cdot}7}{25{\cdot}9} = 2{\cdot}11$
HYBRID	AB	AC	BC
F_1	$\frac{72{\cdot}2}{32{\cdot}5} = 2{\cdot}22$	$\frac{66{\cdot}1}{30{\cdot}3} = 2{\cdot}18$	$\frac{69{\cdot}7}{26{\cdot}4} = 2{\cdot}64$
F_2	$\frac{57{\cdot}8}{28{\cdot}9} = 2{\cdot}00$	$\frac{41{\cdot}9}{25{\cdot}9} = 1{\cdot}62$	$\frac{43{\cdot}0}{29{\cdot}8} = 1{\cdot}44$
D. willistoni			
POPULATION	A	B	C
Parental	$\frac{58{\cdot}0}{35{\cdot}8} = 1{\cdot}62$	$\frac{63{\cdot}2}{33{\cdot}5} = 1{\cdot}89$	$\frac{58{\cdot}5}{26{\cdot}8} = 2{\cdot}18$
HYBRID	AB	AC	BC
F_1	$\frac{62{\cdot}5}{30{\cdot}8} = 2{\cdot}03$	$\frac{69{\cdot}7}{36{\cdot}9} = 1{\cdot}89$	$\frac{72{\cdot}8}{33{\cdot}6} = 2{\cdot}17$
F_2	$\frac{41{\cdot}9}{37{\cdot}2} = 1{\cdot}13$	$\frac{51{\cdot}8}{38{\cdot}8} = 1{\cdot}34$	$\frac{60{\cdot}8}{29{\cdot}2} = 2{\cdot}08$
D. paulistorum			
POPULATION	A	B	C
Parental	$\frac{56{\cdot}7}{33{\cdot}3} = 1{\cdot}70$	$\frac{54{\cdot}4}{36{\cdot}2} = 1{\cdot}50$	$\frac{45{\cdot}6}{30{\cdot}9} = 1{\cdot}48$
HYBRID	AB	AC	BC
F_1	$\frac{54{\cdot}2}{35{\cdot}3} = 1{\cdot}54$	$\frac{50{\cdot}2}{30{\cdot}3} = 1{\cdot}66$	$\frac{64{\cdot}4}{41{\cdot}8} = 1{\cdot}54$
F_2	$\frac{55{\cdot}3}{38{\cdot}1} = 1{\cdot}45$	$\frac{43{\cdot}8}{33{\cdot}2} = 1{\cdot}32$	$\frac{52{\cdot}2}{31{\cdot}6} = 1{\cdot}65$

Vetukhiv measured fecundity by counting the mean number of eggs laid by a female during a 24-hour period, and also as the total number of eggs laid in a lifetime. He used *D. pseudoobscura* and obtained essentially the same results with both measures in four different populations. All 10 possible inter-population crosses were performed, and in each case he compared the female with the two parental strains. Eight of the hybrid females showed a superior fecundity to both parental types, when he measured the number of eggs per day, and seven when he compared the number of eggs per lifetime. This advantage was not maintained in the F_2, where several of the females were inferior to both parents.

Wallace

The results of an essentially similar experiment to the last have been published by Wallace.[265] He used *D. melanogaster*, and measured both the rate of development and the survival under crowded conditions. The experimental flies either came from a single population or were direct F_1 hybrids between two such populations. In addition he was able to examine flies that contained a pre-determined number of chromosomes produced by inter-population recombination. The details of his methodology are complex and need not concern us here. The results, however, show that the greatest number of survivors emanated from the F_1 inter-population hybrids. The survival rate was progressively reduced as the proportion of inter-population recombinant chromosomes was increased (see Table 7.3).

Table 7.3 A comparison of the fitness of flies that have varying proportions of their loci carrying alleles from different populations ('heterozygous') and varying proportions of their chromosomes being recombinant products of different populations. The lines connect classes that differ statistically at the 5% level. Note that the fitnesses are lowest when there has been most recombination, despite the fact that the amount of heterozygosity may be the greatest possible at 100%. From Wallace.[265]

	Recombinant chromosomes	'Heterozygous' Loci		
		100%	50%	0%
(a)	0%	154·38 ± 2·34		148·13 ± 2·91
	50%	140·06 ± 2·20	142·36 ± 2·17	
	100%	140·58 ± 3·96	133·42 ± 3·53	
(b)	0%	171·80 ± 1·94		156·00 ± 2·95
	50%	165·30 ± 1·17	161·00 ± 1·52	
	100%	146·73 ± 1·61		

(a) Mean number of flies present in vials at 11th day.
(b) Mean number of flies in final count of vials.

Wallace's results tell us more than this, however. He also took flies that originated from four different populations (say A, B, C and D), and produced hybrids between them to give AB and CD individuals. Recombination was allowed, and flies were produced that contained one haploid complement from the AB recombinant stock, and one from CD. These flies had the lowest viability of all, which suggests that the physical breaking of gene complexes has a more radical effect upon fitness than hybridization alone. At every locus in its genome, the F_1 hybrid contains two alleles, each from a different population. So do Wallace's complex flies. The difference lies in the latter having bits of 4 gene pools in their complement, where standard F_1 flies have an entire haploid set from each of two. The coadapted complexes which have evolved in the ancestral populations have been broken up by recombination, and the components of fitness are reduced accordingly.

A complication

It would be satisfying if the experimental studies on coadaptation ended here with a neat and consistent set of results. However, an equally careful study by MacFarquhar and Robertson[179] gives contradictory results. They also undertook a detailed programme of crosses within and between populations using *D. subobscura*. The parameters that they measured were body size, development time, and survival at a variety of temperatures and on different kinds of food. The F_1 flies were intermediate between the parental stocks, even when they came from populations as widely separate as Scotland and Israel. Furthermore, there was no evidence of a breakdown in performance in the F_2, nor was there an increase in the variance of these flies as might be expected if coadapted and stable complexes were disintegrating.

These results are in marked contrast to those of Brncic, Vetukhiv and Wallace (and others), and the difference must be accounted for if we are to progress in our understanding of the complexities of genetic interactions within populations. MacFarquhar and Robertson point out there are some weaknesses in the experimental design of Vetukhiv and Brncic. Their flies had been maintained in the laboratory for a variable number of generations prior to experimentation. They will consequently have been subjected to an indeterminate amount of inbreeding, in addition to selection for survival under laboratory conditions. However, this may not be relevant to the differences between inter- and intra-population crosses, for all the stocks will have been subject to the same forces.

At present it does not seem possible to resolve the conundrum.

Drosophila subobscura does not show the predicted results of coadaptation, possibly because the species does not show coadaptation. *D. pseudoobscura*, *D. willistoni* and *D. melanogaster* all show some evidence that their populations are coadapted, but we cannot assume the generality of the phenomenon from these inconsistent results. Fortunately, the study of coadaptation has been continued elsewhere, using different material and a different approach.

THE EVOLUTION OF DOMINANCE

An important array of evidence pertaining to the coadapted nature of genetic material stems from studies into the dominance relationships of certain alleles. It is not proposed to discuss the phenomenon of dominance in detail since it is a complex subject which has been well reviewed by Murray[201] and is too involved for an introductory text of this nature. We shall briefly review the present situation, reporting and discussing some studies which cast particular light upon the problem of coadaptation and the interactive nature of loci.

In 1928, R. A. Fisher proposed that dominance must be a modifiable parameter, since experimentally derived mutants are usually recessive. There seemed to him to be no reason why this should be so, and yet, as a general rule, structural mutants were recessive to the wild type. He argued[94–5] that newly-arisen mutations should be intermediate, but that modifying genes at other loci affected and changed the performance of the ancestral genes, so that heterozygotes manifested the original phenotype. Then, dominance evolved during the early life of a mutation.

In support of his theory, Fisher cited studies on the mutant gene 'eyeless' of *Drosophila melanogaster*. Flies that are homozygous for this gene show a marked reduction in the size of the eyes. If a line is established which is fixed for the 'eyeless' gene, over the course of time, the expression of the character gradually changes back towards wild type. However, at any generation, if the line is crossed to another stock, there is a reversion to the reduced eye phenotype in homozygotes. It seems as though there is a reorganization of the gene pool in the line to maximize the size of the eyes. It is easy to see how this could happen, for vision is an important component of courtship in *D. melanogaster*, and those flies with bigger eyes will be more successful in obtaining mates. The constellation of alleles that they carry will thus be over-represented in the next generation. Any genes that act to modify the size of the eyes will consequently increase in frequency. Gradually, a gene-complex will be established which optimizes the size of the eyes in flies that are homozygous for the 'eyeless' gene. Out-crossing these flies to another population will destroy this gene-complex and so the eyes will revert to the original 'eyeless' size.

Modification of dominance in this way has also been reported by Kettlewell[146] in the moth *Biston betularia*. He compared specimens of the *carbonaria* melanic collected in the nineteenth century with contemporary material, and found that melanism was more complete in the latter. The early material was marked with pale lines which are not present today.

These ideas on evolving dominance had been challenged by Wright[274, 275] partly upon the grounds that the evolution would be far too slow in the wild. He analysed the evolution of dominance under the effect of genes, at a separate locus, that modify the phenotype. His theoretical results suggested that the changes in gene-frequency at the modifier locus would be similar to the mutation rate, and consequently be likely to suffer random fluctuations due to inbreeding or drift. He agreed that once the alleles had become sufficiently common, then selection would proceed rapidly, but considered it unlikely ever to get off the ground in the first place.

Wright also suggested that dominance is a property of the alleles themselves. Whichever allele is the more active will determine the properties and performance of the heterozygote. This is most obvious in the case of certain enzymic genes. A particular allele may be biochemically inactive, perhaps because of a substitution close to the active site. A heterozygote will then bear the qualitative phenotype of the active allele which is present. The quantitative phenotype will depend upon the biochemical properties of the enzymes—their rates of reaction, stabilities and so on. This is the basis of Haldane's[119] views on the subject of dominance, which are also somewhat at variance with those of Fisher and his supporters.

The argument has ebbed and flowed over the years, but the basic observations remain. Firstly, dominance may increase over the generations, as with *B. betularia*. Secondly, it is possible to improve dominance by artificial selection. Ford[101] showed this with *Abraxus grossulariata*. He crossed typical white moths to the cream coloured *lutea* heterozygotes. In one series of matings Ford chose the darkest heterozygotes as parent in the next generation, in another he took the palest. In a few generations, he produced heterozygotes which were indistinguishable from the typical whites and others which were phenotypically the same as *lutea* homozygotes. It seems that his original stock contained sufficient modifying genes to allow dominance to be swiftly altered in either direction.

The results suggest that (at least some) loci carry genes that are subject to the action of modifier loci. It is selection upon these modifier loci that is responsible for changing the dominance relationships, and when dominance has evolved the loci involved can be regarded as a coadapted gene-complex. We can make a prediction from this. Suppose an allele is present and shows dominance in one population. If there is another population which is geographically isolated from the first and which does not carry the allele, the modifying loci in the latter should not have been subjected to selection to improve the concordance between homozygote and heterozygote. The former population should include a complex of genes which act to produce dominance, and crossing individuals from the two populations will disrupt this complex. Dominance should therefore break down in the hybrids or their derivatives.

Kettlewell[147] has performed just such an experiment, again using *Biston betularia*. In western Europe, *carbonaria* is totally dominant, and Kettlewell believes this to have arisen because predators are sufficiently discriminating to be able to detect the reduced crypsis of the paler-patterned heterozygotes. He crossed these *carbonaria* individuals with typical moths of the extremely closely related 'species' *B. cognataria* from Canada where melanics are virtually unknown. The pattern of dominance broke down after only two or three generations, and heterozygotes became more intermediate.

A similar result has been provided by the detailed analysis of mimetic polymorphism in the genus *Papilio* by Clarke and Sheppard. There is a species called *P. dardanus* which inhabits much of tropical Africa. It resembles *P. memnon*, which we discussed earlier, in that the males are constant in phenotype and non-mimetic, while the females are polymorphic and often have several mimetic forms in a single population.

There are several races of this insect from different parts of Africa. In some the females are also non-mimetic. Conversely, the same mimetic form may be found in a series of different races. Thus, the variaty called *cenea* mimics the distasteful *Amauris echeria* and *A. albinaculata*, and is found in the races Cenea from South Africa (85%), Meseres from Lake Victoria (7%) and Polytrophus from Northern Kenya (6%). It is however absent from the race Dardanus found in West Africa. Clarke and Sheppard[50–3] crossed female *cenea* from South Africa with males from all four of these races. In the South Africa cross, *cenea* behaved as a good autosomal dominant supergene, and the offspring which carried it were good mimics of the distasteful *Amauris* models. Hybrid butterflies formed by crossing *cenea* with Meseres or Polytrophus were less perfect mimics, and those from Dardanus showed little resemblance at all. These results suggest that modifying genes are present in the South African populations which adjust the manifestation of the *cenea* supergene to produce a high-quality mimic. In populations where *cenea* is naturally rare, the modifying genes are less well adapted, and so the mimetic resemblance is reduced. Populations which lack *cenea* have not been subjected to selection to improve the mimicry at all, and so the resemblance is almost non-existent.

Thus our prediction is borne out. The action of avian predators upon the mimetic complex has produced a constellation of alleles at the modifying loci which confer an increased resemblance upon the mimic. These alleles form a coadapted complex because enhanced fitness depends upon the possession of sufficient modifying alleles to promote close mimicry. If the butterfly differs from its model, it will be recognized as distinct and eaten.

AREA EFFECTS

In the Mojave Desert of California, *Linanthus parryae* is to be found. This is an insect-pollinated, self-incompatible annual plant which in some years is so abundant that it carpets the desert, yet in other years it is very thin upon the ground. It is polymorphic for colour in the western Mojave Desert, occurring either with blue or white flowers. Populations in this area may be all white, or all blue, as well as polymorphic. In an early study of the population genetics of this species, Epling and Dobzhansky[88] ascribed the apparently haphazard distribution of blue and white flowered plants to random genetic drift.

A reappraisal of their data by Wright[278] suggested that natural selection was involved, and more recent results by Epling, Lewis and Ball[89] have confirmed this. They followed a series of populations for over twenty years (generations). They found that the relative frequencies of blue and white flowers remained constant in a population despite large fluctuations in the population density, and presumably, therefore, population size. They were unable, however, to determine the selective agents acting upon the flower colour genes to maintain the phenotype frequencies.

A possible explanation of this situation is that the selection does not act upon the flower colour at all. The amount of gene flow between populations is small, and the population sizes are low,[89] so local differentiation of the populations could have taken place to produce coadapted gene complexes each of which has its own particular array of flower colour alleles. The long-term stability of flower colour might thus be explained. However, if there are micro-geographical coadapted races, as this hypothesis must imply, they do not appear to be associated in any consistent way with the local microclimate as reflected by the habitat.

A similar phenomenon has been described by Cain and Currey[27, 28] in the land snail *Cepaea nemoralis*. On areas of high land in southern England, they found that the pattern of visual selection first described by Cain and Sheppard[31, 32] broke down. As with *Linanthus parryae*, phenotypes were present at a relatively constant frequency over areas that are large with respect to the population size. These patterns were constant over a variety of backgrounds, and then changed quite suddenly—often in the middle of an apparently uniform habitat (see Fig. 7.7). Cain and Currey called these patterns of distribution 'area effects', and similar phenomena have subsequently been described from several other species of land snails. For example, Clarke and Murray[47] have described area effects for shell-character morphs in species of *Partula* from the island of Moorea in French Polynesia. Here again, there is no evidence of any environmental factor being associated with the observed patterns of distribution.

The maintenance of these area effects in *Cepaea nemoralis* has been a matter of considerable dispute. It is generally agreed that the area over which the effects extend, and the number of individuals involved, is too great for random genetic drift to be of any major importance at the present time. Cain and Currey[28] were able to show from sub-fossil material that some of the effects had probably been present for several thousand years. Furthermore, there was no evidence of a reduction of the habitat suitable for *Cepaea*, either by climatic crisis or changed agricultural practice. Thus, it was not possible to argue that the populations had been reduced to refuges within which they had evolved balanced gene complexes, prior to an amelioration of conditions, a spread of populations, and secondary contact of these stable complexes. The answer had to lie elsewhere.

Goodhart[111] postulated that the area effects on the Malborough Downs reflected the structure of the colonies of *C. nemoralis* that originally founded the populations there. The differences between these founding populations persisted through the evolution of coadapted complexes. Mixing along the zones of contact was prevented by an inferior fitness of the hybrids. Wright[280] put forward essentially the same argument, suggesting that the clines that were produced would be stable if each coadapted complex were best adapted to its own area.

Cain and Currey[29] were unimpressed by Goodhart's argument, and favoured a purely selectionist approach. They pointed out that the complexes were unlikely to be equal in fitness and one or another

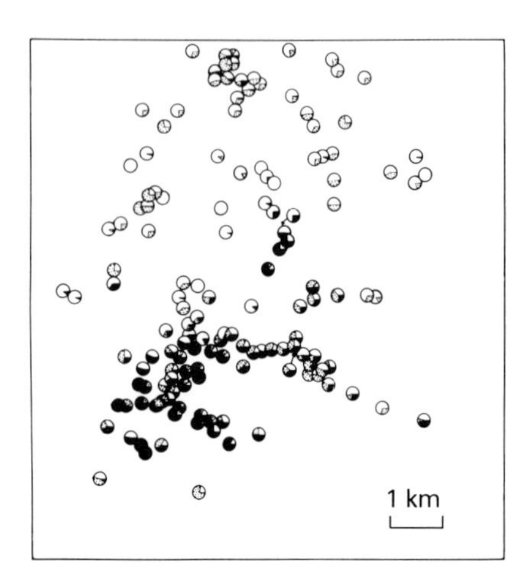

Fig. 7.7 The proportions of yellow (white), pink (stippled) and brown shells in a series of samples of *Cepaea nemoralis* taken from the Marlborough Downs in southern England. Note the 'area effect' pattern of high frequencies of a phenotype changing rapidly and over a short distance to another. The pattern bears no relation to background vegetation. From Cain and Currey.[28]

should slowly but steadily replace all others. Furthermore, the pattern of area effects itself did not fit with a founder effect hypothesis. A comparison of Figs 7.7 and 7.8 makes it clear that the boundaries of the area effects for the closely linked loci for colour and banding do not coincide. The implication must be that different effects are operating upon the two loci, since arguments based upon founding populations would predict concordance rather than difference.

Clarke[40] has put forward stronger evidence in support of a coadaptational basis to area effects in snails. He describes a part of the Polynesian island of Moorea where the populations of *Partula taeniata* are polymorphic for shell colour. Over much of the north-western Moorea, the populations manifest a low level of purple-shelled individuals, but in one area the frequency rises to about 50%. Clarke reports upon a transect of 20 samples across the boundary from low to high purple. *Partula* is a relatively immobile snail that spends its inactive hours above ground on trees and vegetation. Consequently, it is easily collected, and sample size reflects the population size rather closely. Figure 7.9 shows the results of Clarke's transect. The frequency of purple drops steeply over about 70 metres, most rapidly in an area where the population is large. It seems

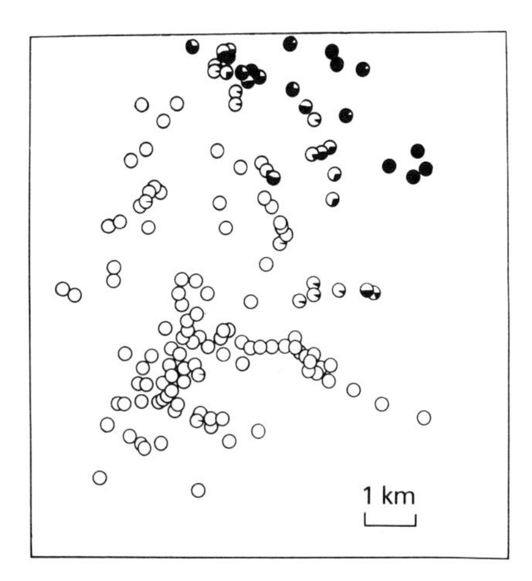

Fig. 7.8 The proportions of five-banded (black) and non-five-banded (white) shells in the same samples of *C. nemoralis* from the Marlborough Downs. Note the lack of concordance with Fig. 7.7, suggesting that the two distributions are not the results of similar historical events.

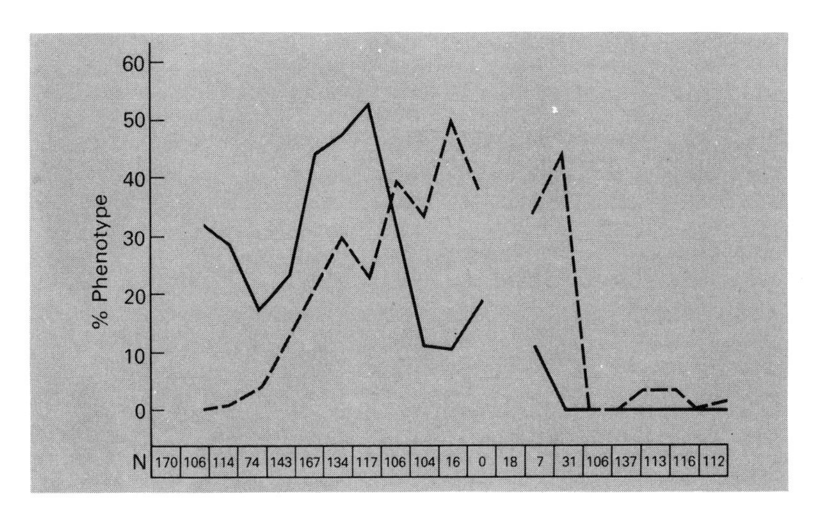

Fig. 7.9 The percentage of purple shells (———) and banded shells (– – – – –) in samples of *Partula taeniata* taken from twenty adjacent 10 m squares. The transect ran from an area of high purple to one of low. The boxes give the numbers of adult snails collected. From Clarke.[40]

unlikely, then, that the change in the frequency of purple is related to any geographical barrier to the movement of the snails.

It is very interesting to note that, in the middle of the transect, where the change in frequency of purple is strongest, there is an increase in the abundance of banded shells. The frequency reaches 50% in the middle, which is far above that in any other part of north-west Moorea[40]. Furthermore, the *expression* of banded in this transect is less complete than in other populations. Clarke considers that this may be due to special conditions in the genetic environment—perhaps resulting in a partial breakdown of dominance. In addition, in the same region of the transect there are marked heterogeneities in the mean shell size and fecundity of the populations. It is possible that all of these results stem from a breakdown in the genetic structure of the populations along a zone of hybridization between two coadapted complexes.

Clarke then proceeds to argue that these complexes may have arisen in the absence of migration barriers if some of the loci are subjected to strong selection. Genes that are compatible with the favoured alleles will also increase in frequency, with steep clines forming along the boundaries. Over the course of time, the populations will evolve into a series of coadapted complexes with more or less sharp changes of gene frequency between them. The patterns of area effects which are observed in *Linanthus, Cepaea, Partula*, and other species, might then be a reflection of these complexes. Migration between adjacent populations will produce clines whose steepness will depend upon the mobility of the species, and the balance of advantage and disadvantage of individuals from each complex in their own or the alternative environment.

8

Some Ecological and Population Genetics Aspects of Trans-specific Evolution

INTRODUCTION

Charles Darwin's great treatise *On the Origin of Species* was published in 1859, and it is clear that, at that time, its author inclined to the view that a species was little more than a well-marked and clearly defined variety. He accepted that it might be of a fairly permanent nature, but considered that species in general were essentially artificial groupings, of no greater biological reality than orders or genera. Yet he clearly believed that biological classification itself was not arbitrary: it could and should reflect quite precisely the patterns of descent and evolution of all living things.

The years that have followed Darwin's early steps have seen a close and careful scrutiny of the nature of species. This has resulted in a gradual crystallization of ideas pertaining to this systematic category and its relationship to the rest. It is now generally agreed that supra-specific taxonomic characters are pretty subjective. Objectivity is the desired goal, but is difficult to attain because it is a matter for the taxonomist's opinion and judgement whether a group of animals or plants is placed in an infra-order or an order, a family or a sub-family. His fellow taxonomists may agree or differ, and can argue and discuss the finer points of classification with him. Their opinions, however, are as subjective as his own. Attempts to refine and standardize procedures have resulted in the development of techniques in numerical taxonomy. This discipline attempts to provide objectivity by algebraically manipulating a large series of comparisons simultaneously, but the decisions on whether to include or omit characters from the analysis remain subjective.

The careful appraisal of the significance of the species category that we alluded to above has resulted in much less dispute as to its nature. It is almost universally accepted to be a real biological entity, permitting precise definition. The most widely accepted definition is that of Ernst Mayr, who suggested[182] that species are: 'groups of actively or potentially interbreeding natural populations which are reproductively isolated from other such groups.' The essence of this definition is that a species is an interbreeding unit, members of which cannot breed successfully with members of another breeding unit. Inability to breed successfully may, of course, reflect not only genetic barriers, but also physical or temporal ones. We belong to the same species as our ancestors of 200 years ago, even though we cannot interbreed with them. As we shall see, however, such temporal isolation can lead, over much longer periods of time, to the appearance of what, in the judgement of palaeontologists, are new species.

Species, then, are taxonomic groupings that are capable of clear definition and have a biological reality. A group of animals or plants, or even a single individual, can usually be assigned to one or other of the million and more species that are now known to science. These species themselves can be arranged into assemblages on the basis of greater or lesser resemblances. The evolutionist believes that this has a deep biological significance, reflecting to a large extent the history through which the species have passed. We shall devote the remainder of this chapter to an assessment of the contribution that ecological and population genetics have made to our understanding of the mechanism of this evolutionary process, starting with a brief consideration of the genetic differences between species themselves.

Back in 1901, during the dawn of genetics, De Vries suggested that one species could arise from another as the result of a single mutation that prevented hybridization. This belief was strongly supported by Goldschmidt[110]—and is even now sometimes echoed in student essays. However, apart from a few specialized instances among unicellular organisms, there is no evidence that a mutation at a single locus has ever given rise to specific distinction between two groups of plants or animals.

Nevertheless, novel species can arise as the result of a major genetic reorganization in a single individual, but the cause is more drastic than a single simple mutation. Species can be found that have two or more times the number of chromosomes of another. Such a phenomenon is called polyploidy, and is very common among plants. We shall give it no more than passing reference here; it is fully discussed by Stebbins[250] and other authors.

There remain, however, about 50% of plant species, and the

overwhelming majority of animal species, that have not arisen by chromosomal multiplication. It seems that the differences between them have arisen slowly and imperceptibly by the accumulation of mutations over long periods of time, either by random events or through the process of natural selection. Since ecological and population genetics have done so much to uncover the mechanisms of evolutionary change in natural populations, we shall proceed to an examination of their role in helping the elucidation of speciation by this process of gene substitution.

ANAGENESIS AND CLADOGENESIS

Examination and analysis of the fossil record repeatedly shows that groups of organisms change and evolve into new forms as time passes. The earliest mammals, for example, appear in the Rhoetian layers from the boundary between the Triassic and Jurassic eras, about 150 million years ago. The relics that have been found are all small, and mostly consist of teeth, but examination of later material clearly shows that their affinities lie with the mammals. By piecing together remains from different localities and different geological eras, a pattern emerges of the evolution and increasing complexity of the mammalian form, and of particular groups within that lineage.

Frequently, material is missing from the sequences, and so the detail of these evolutionary transitions goes unrecorded. Occasionally, however, large numbers of specimens are available, so that the changes in form and structure can be recorded almost completely. Perhaps the best and most familiar example concerns the history of the Equidae, during which the horse has evolved from a small, dog-sized animal that existed in the early Eocene. In this lineage, the sequences are so complete that the power of natural selection to develop a free-ranging herbivorous animal from the primitive early Perissodactyl material is plain to see.

The earliest horses had four toes which ended in a tough, flexible pad of tissue.[242] Gradually the central toe lengthened and strengthened (Fig. 8.1) taking ever more of the weight of the animal. The fourth toe was lost, and later the first and third as well, to give the single-toed animal that we know today as *Equus*. Equally important with the reduction in toe number was the development along and around the bones of strong ligaments, which acted as a spring to speed the animal on its way, and could only evolve when horses stood upon their toes. These adaptations served to strengthen and lengthen the lower limb, thereby allowing greater speed to escape predators, and a

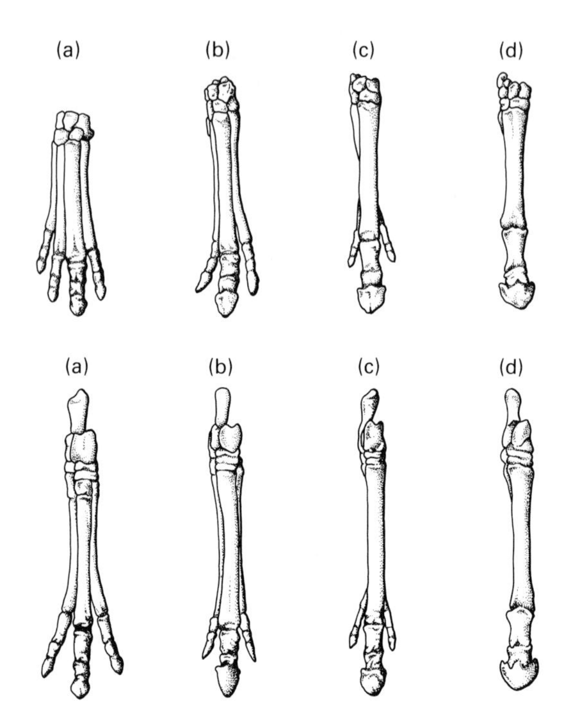

Fig. 8.1 The bones of the lower limbs of horses, anterior above and posterior below. (**a**) *Eohippus*, a primitive lower Eocene animal. (**b**) *Miohippus*, from the Oligocene. (**c**) *Merychippus*, from the late Miocene. (**d**) *Equus*, the modern horse. Note that the bones have fused together, and become relatively longer and more slender, to give height, lightness and strength. From Romer (1966). *Vertebrate Palaeotology.* 3rd edition. Chicago University Press.

high mobility to search for food on the savannah plains that were the horses' habitat.

The bare outline depicted in Fig. 8.1 can be extended and expanded by the inclusion of many other fossil forms. The eventual picture of equid evolution (Fig. 8.2) shows there to have been a series of different evolutionary paths. Some of these seem to have been straight and relatively direct, others are branching and complex. All, however, end in extinction, save only the lineage leading to *Equus* itself.

The term anagenesis was coined by Rensch[225] to describe the evolutionary pattern where one form develops more or less directly into the next. An example from Fig. 8.2 is the line leading from *Hyracotherium* through *Orohippus* and *Epihippus* to *Mesohippus*. The branching pattern as shown by an early form of *Archytherium*

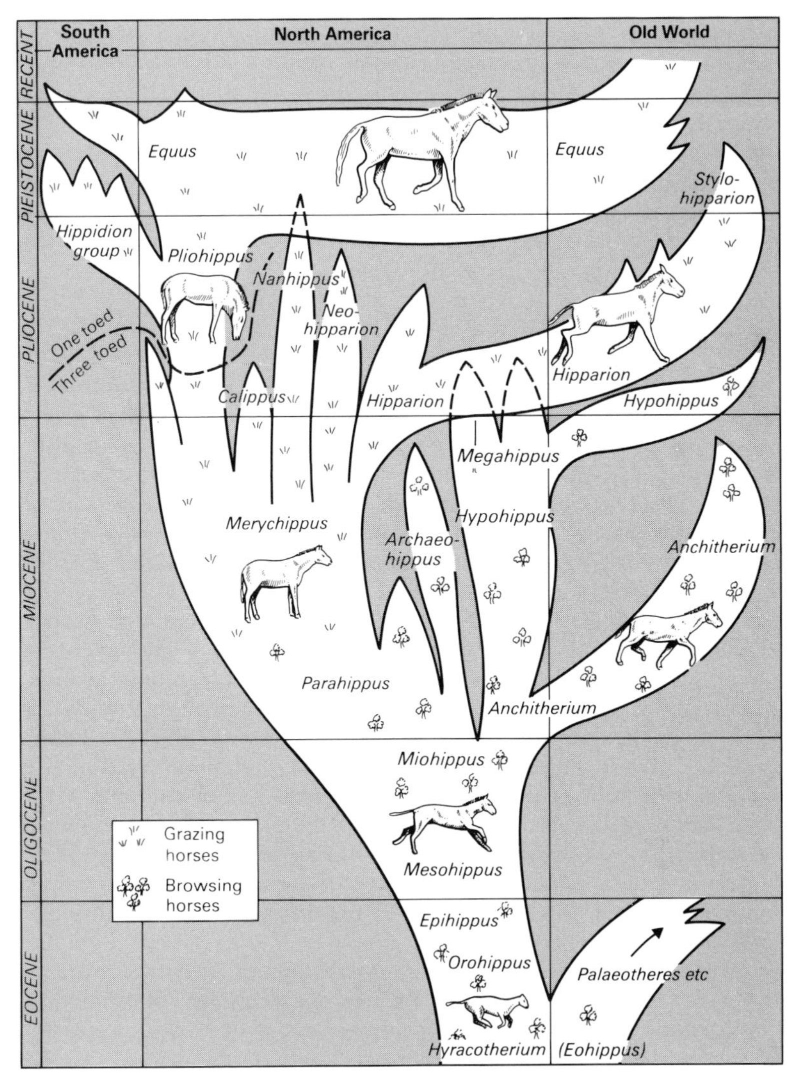

Fig. 8.2 The evolutionary lineage leading to the modern horse, *Equus*, from the Eocene *Hyracotherium*. From Simpson.[242]

splitting into *Hypohippus* and a more advanced form of *Archytherium*, Rensch termed cladogenesis.

While these terms of Rensch have to some extent fallen into disuse in recent years, they do serve to draw our attention to two fairly distinct kinds of speciation. The former of these is the progressive development of a species through time, changing and evolving from one form into another. The other is the method by which species increase in number. One form gives rise to another additional to itself—usually following the physical partitioning of the populations into two groups. These then evolve along separate paths, finally accumulating sufficient differences to be genetically distinct. We shall now consider these two kind of speciation in turn, and discuss the light cast upon them by population genetics.

EVOLUTION THROUGH TIME

Another fine example of anagenetic and cladogenetic evolution is to be found in the irregular echinoids of the genus *Micraster*. This group is now extinct, but the fossil remains of their shells or tests can be found in considerable numbers in the chalk of localities such as Beachy Head in southern England. These chalk deposits can be divided into a series of horizontal zones or bands on the basis of their faunal composition, and these zones in turn can be approximately related to the time at which the chalk was deposited.

A. W. Rowe collected *Micraster* from the chalk cliffs of Beachy Head, carefully recording the zonal position of each specimen. Although there is a considerable amount of variation among the fossil *Micraster* from any individual zone, clearly discernible trends are apparent in the morphology of the test from layer to layer. Ascending from the lowest and oldest zone, it becomes relatively broader. In parallel with this, the positions of the broadest and tallest parts of the shell move posteriorly, the mouth moves anteriorly and the ambulacral areas become progressively more ornamented.[205] However, at all levels there is a considerable amount of variation, and individuals have been found that show many of the characteristics of earlier or later forms. Nevertheless, there seems overall to have been a gradual evolutionary trend, changing the morphology of *Micraster* through the ages. As with the horses, we can see examples of branching cladogenetic evolution, and the anagenetic progression from one species to the next.

Thus Rowe,[231] Imbrie,[128] Nicholls[205] and the other *Micraster* taxonomists consider that the array of forms can be divided into five species (Fig. 8.3). The oldest of these is probably *M. leskei*, which gave rise to both *M. cor-bovis* and *M. cor-testudinarum*. The latter of these was the ancestor of both *M. senonensis* and *M. cor-anguinium*, probably as shown in Fig. 8.3. It is impossible to be certain now that these forms, especially those on the main lineage, were truly specifi-

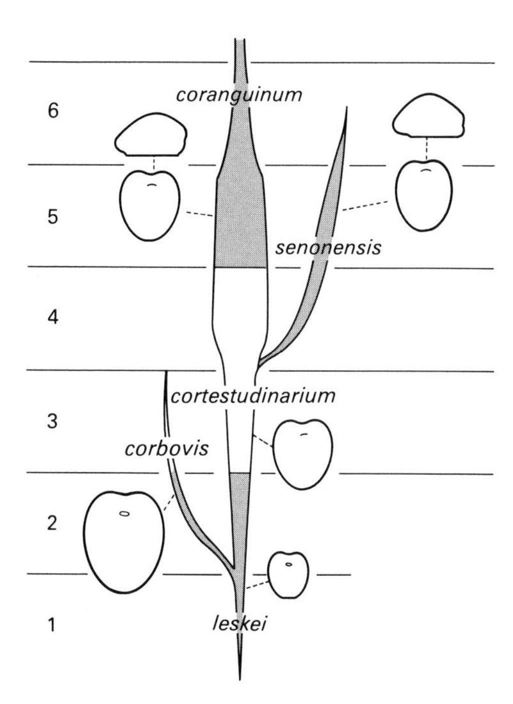

Fig. 8.3 Postulated evolutionary relationships of five forms of the echiniaid *Micraster* through six geological time periods in the chalk of southern England. From Imbrie.[128]

cally distinct, in the sense of Mayr's definition, and not merely races or sub-species of a single variable species. The fact that some forms coexisted without apparent hybrid forms does, however, suggest that they were to some extent reproductively isolated from one another.

How then do we explain this evolutionary procession through time? It seems highly unlikely that environments remain static, stable and unchanging over prolonged periods of time. By the very nature of things, their physical and biological components are in a constant state of flux and change, either slowly or rapidly, but nevertheless altering the environment in which a species lives. For the population to survive under these conditions, it must continually evolve and change in response to the ebb and flow of forces acting upon the biological community of which it forms a part. Periodically, new alleles will arise by mutation, while inversions and translocations will disrupt and reorganize the genome, continually throwing up new combinations of alleles within the genetic architecture of the population or species. Random processes may lead to the spread of some of these novelties, while natural selection will cause the increase of some of those that confer an advantage upon their bearers. Slowly, the species will evolve.

The preservation of this continuous and progressive evolution in the fossil record has both advantages and disadvantages for the student of evolutionary genetics. On the one hand, it is possible for him to compare specimens directly, and to observe and measure the changes that have taken place. On the other hand, it is not possible to analyse living members of the series, and thereby directly determine their genetic relationships. In principle, when two individuals are both extant, it is possible to attempt to cross them, and so the identification of specific differences in living material might be thought to be straightforward. In practice, though, it is far from this, as volumes of literature bear mute testimony. However, when the only material available is fossilized, judging what may or may not be sufficient differences to allow the allocation of specific rank remains a matter for a taxonomists's judgement. Hence the suggestion of Cain[26] that 'a species is a species if a competent systematist says it is.'

ISOLATION BY DISTANCE

In the evolutionary lineage of *Micraster*, each generation is connected to the temporally adjacent one, and we may suppose that each responds slightly to any change in the environment. As we have noted, the change in appearance of the populations composing the species may be quite imperceptible in the short term, and only becomes apparent when we compare individuals that are separated by substantial periods of time.

However, progressive changes in environmental parameters do not only occur with the passage of time. Changes in geographical position due to altitude, longitude or latitude can also be reflected in regular changes in the environment. We have already discussed how this affects *Colias* butterflies in western North America.[269] Populations inhabiting higher altitudes or more northerly latitudes are phenotypically darker than those from more temperate climes. Similarly, in the New World populations of *Passer domesticus*, populations from the most continental climates comprise larger individuals with relatively shorter extremities than those from warmer situations.[136] Both this and the *Colias* example can be accounted for in terms of thermal adaptation. Each population is at, or evolving towards, the genetic structure that produces optimal phenotypes for its particular habitat.

Migration occurs between adjacent populations, and acts to reduce the individual adaptedness. However, eventually some sort of equilibrium is attained where the selective loss of individuals, and hence alleles, will be balanced by immigration from adjacent populations.

Despite the relatively large differences between sparrow populations in North America, no one has seriously suggested that they are not all of the same species. Nevertheless, it has long been appreciated that a widely distributed species with a relatively low mobility might show differences between the extremes of its range that would deserve specific rank, were the connecting forms not extant.

A good example of this is given by the great tit (*Parus major*). This bird is widely distributed in the Palaearctic, occurring in a giant ring around the Central Asian plateau (see Fig. 8.4). In Europe, *P. major* has a green back and yellow belly, and this phenotype occurs right across northern Asia to the Pacific Ocean. The southern loop of the range shows greater variation, however, with a steady grade or cline of increasing greyness through Iran to a grey-backed and white-bellied form in India. Through the Chan region north of the Malaysian peninsula, the back becomes progressively greener, until the Chinese population are fully green, though still with a white belly. This form overlaps with the yellow-bellied form in the extreme east of Siberia, and the two apparently coexist. It was at one time thought that these two forms behaved as good species, but it is now believed

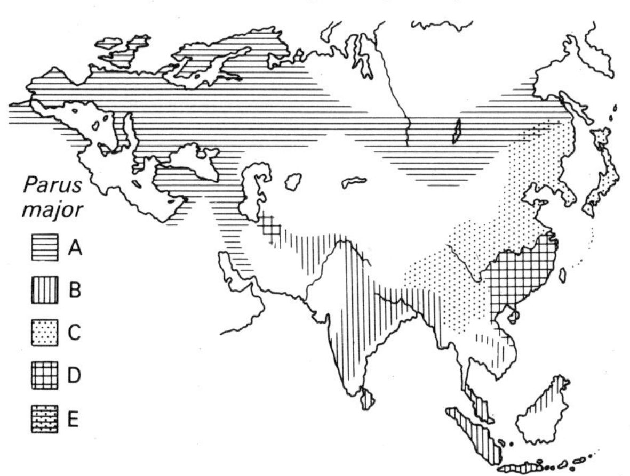

Fig. 8.4 The ring species complex of the great tit, *Parus major.* **A**, green back and yellow belly. **B**, grey back and white belly. **C**, green back and white belly. Hybrid populations are found in area D, and were once believed to be separate species in area E. After Mayr.[183]

that hybridization does occur, but how often or how successfully remains obscure.

Until a detailed field analysis has been made of the breeding biology and ecology of these birds in their area of sympatry, it will not be possible to define the situation more clearly. However, it seems possible that this is an example of what is termed a ring species—so-called because of their circular distribution patterns. Ring species have been reported in gulls of the genus *Larus*,[183] and in a variety of other groups. They probably originate from a localized species that has spread its range around an ecologically unsuitable barrier. As the species has expanded, its populations have evolved to the local conditions, gradually accumulating differences from the parental gene pool. Eventually, the two ends of the range come into contact. If sufficient differences have evolved, then the two forms will behave as good species and a true ring species will have arisen. With less complete differentiation, there will be only partial or incomplete isolation. The true nature of this needs to be analysed at the ecological, behavioural and genetic levels. For example, in the *Larus argentatus* complex of gulls, two forms, *Larus argentatus* and *L. fuscus*, lie at opposite ends of the distribution and are sympatric in the British Isles, but do not normally hybridize in the wild. Rare hybrids have been reported, but in mixed colonies isolation is ecological and behavioural, for the two forms breed at slightly different times and choose marginally different nesting sites.

The confirmation of genetic differences in ring species ideally involves the experimental crossing of the two forms, and the detailed analysis of their genetic structure. Evidence of reproductive failure can often give an indication of incompatibilities associated with the prolonged separate evolution. Such experimentation is almost impossible with birds such as *Parus* or *Larus*, but has been undertaken extensively in another species that was formerly considered to show isolation by distance.

The North American leopard frog (*Rana pipiens*) is found widely through the sub-continent from Canada in the north to Costa Rica in the south. It was formerly believed to be a supreme example of isolation by distance, but recent work demands some reappraisal of this conclusion.

A long series of experiments were performed during the twenty years following 1940 by Moore,[190–4] Riubal[228–9] and Volpe.[262–3] These consisted largely of attempted hybridizations between frogs coming from closely adjacent or widely separated populations. The results seemed to indicate that *Rana pipiens* was a single species composed of a series of adjacent and inter-breeding populations.

Frogs from the same or closely adjacent localities were fully interfertile, producing large numbers of offspring, with relatively little death during development. Crosses between northern and southern frogs, however, showed a high level of incompatibility, manifest in embryonic malformations and mortality. The level of reproductive failure was greater as the latitudinal or altitudinal distance between the parental populations increased. It thus seemed to the earlier workers that *R. pipiens* was one species. It was true that there was evidence of local adaptations (perhaps to temperature), and of interpopulation differences reflecting a slow but progressive change in genetic structure over the range of the species. But, effectively, there seemed to be a single gene pool, extending more or less unbroken from the Caribbean to central Canada.

More recent researchers have, however, cast doubt upon this conclusion. Both Post and Pettus[218] and Cuellar[62] suggest that the interpretation may be erroneous, and that several species may be involved in a patchwork distribution. The more widely separated any two populations may be, the greater is the chance that they are from different species of this complex, and hence the more likely reproductive failure becomes. The evidence in support of this new hypothesis includes behavioural data from the wild, including the analysis of call notes. The situation is not completely resolved, but, as with so many cases of evolution, detailed analysis suggests that the picture is more complex than was initially believed.

ISOLATION BY BARRIER

We have discussed at length the ways in which populations and species can evolve differences over a period of time. It is also evident that if two populations or groups of populations are isolated from one another, they need not evolve along parallel paths. Variations in the physical and biological environments may result in different adaptive strategies being developed in the two isolates. In addition, random processes by their very nature are likely to differ between them, and so these will increase their distinctness. After a sufficient length of time, the differences will be sufficiently great to debar successful hybridization.

The studies of genetic coadaptation of Vetukhiv and others, that we discussed in Chapter 7, have indicated that this might happen to some extent between contemporary populations of some *Drosophila* species. There is no evidence from this research that the populations have yet evolved any major barriers to hybridization, although Prakash[220] has reported that these may be developing in *D. pseudoobscura*.

Populations of this species show comparative uniformity in their protein polymorphism, although an isolated group in Bogota, Colombia, is markedly distinct. Attempts to cross these with flies from further north show evidence of infertility. It seems possible that the isolated population from Bogota may have accumulated a sufficiently distinct constellation of genes that it may be on the verge of specific separation.

Such incipient or nascent speciation can only normally be detected when the organisms are sufficiently convenient to be manipulated experimentally in the laboratory. Separate isolated populations are rarely observed to come into secondary contact, and so the interpretation of their taxonomic status becomes a matter of judgement. There are exceptions to this rule, however, particularly among the more conspicuous and better studied groups. An especially informative series of examples can be found in the effects of glacial ice upon several European passerine birds.

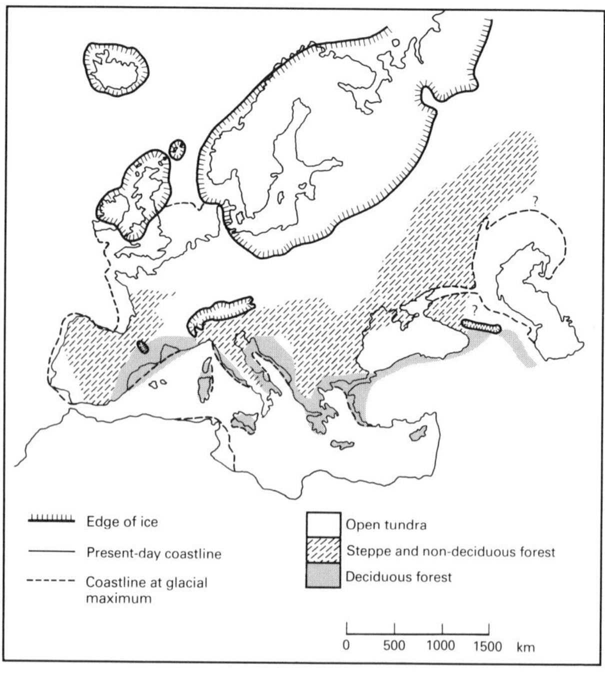

Fig. 8.5 The vegetation zones of Europe during the last glacial maximum. The areas of deciduous woodland are restricted to the Mediterranean, and largely to Iberia in the east, and the Balkans in the west. From Moreau.[196]

Eighteen thousand years ago, the geography of Europe was probably somewhat like Fig. 8.5. Moreau[196] suggested that there was probably little broad-leafed forest anywhere in the continent at this time, apart from Iberia, Italy, the Balkans, and parts of Anatolia. We may suppose, therefore, that there was a dramatic reduction in the number of woodland birds brought about by the onset of glaciation, and that the majority of them would have been restricted to these regions. A similar pattern of repeated contraction and expansion of numbers probably occurred in most of the glaciations. Contemporary distributional data give evidence of the consequences of this, particularly relating to allopatric speciation, and we will briefly consider three examples.

The first concerns two closely-related species of warbler, *Hippolais icterina* and *H. polyglotta*. These two forms are very similar in phenotype, but apparently do not hybridize. They thus behave as 'good' species, and show a marked pattern of allopatric distribution (Fig. 8.6). Both species are trans-Saharan migrants, with widely

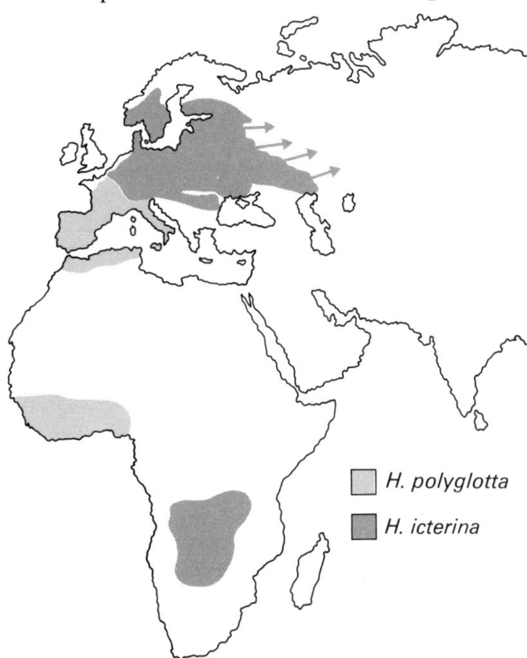

Fig. 8.6 The distribution of two closely related European warblers; *Hippolais icterina* and *H. polyglotta* are allopatric in Europe on the breeding grounds, and spend the winter in widely separate areas of Africa. After Lack.[167]

separate wintering areas, although in the breeding season they come into close contact in Western Europe, with a very narrow zone of overlap. Territorial behaviour is very strong in *Hippolais* warblers, and males will drive other males of either species out of their territories.

The two European nightingales, *Luscinia luscinia* and *L. megarhynchos*, both inhabit lush, bushy areas close to water. They are broadly separate geographically in both summer and winter, although there is some overlap in eastern Europe. Where they coexist, *L. luscinia* occupies the wetter habitats. Contemporary data suggest that *L. megarhynchos* is on the retreat in these areas, being pushed first into drier, less optimal sites, and finally eastward out of parts of its earlier range.

There are two tree-creepers in Europe, *Certhia familiaris* and *C. brachydactyla*. The former is more northerly in distribution, but there is a broad area of overlap in middle latitudes. In this region, *C. familiaris* tends to occupy the coniferous forests at higher altitudes, whereas *C. brachydactyla* is more typically found lower down in broad-leafed woodland. In the south and west, *C. brachydactyla* is found in all woodlands at lower and higher altitudes. There is thus a restriction in the habitat of both species where they overlap.

Such pairs of morphologically similar or identical natural populations that are reproductively isolated from one another are called sibling species. In all of these examples, the presence of two sibling species can be attributed to the effects of prolonged separation by glacial ice. The marked reduction in the number of individuals in each of the refuges would play an initial part in differentiating the genetic structure of the isolates. Subsequent evolution would be along different paths, for the physical and ecological environments would differ in the two main refuges of Iberia and the Balkans. Gene flow would probably cease completely, for even the migrant species seem to have had different wintering areas in central Africa.

Following the climatic amelioration, the separate forms spread northward with the expanding woodland habitat, until once again they made contact. In all three examples, the isolates had evolved specific differences so that hybridization was impossible in the wild. However, the ecologies remained sufficiently close for competition to occur between the sibling forms. In the case of the tree creepers competition was minimized by the two species occupying different altitudinal zones, and developing an apparently stable relationship. The *Hippolais* warblers do not show habitat separation, but maintain an apparently stable boundary between themselves, with little or no overlap. The nightingales are different again, for *L. luscinia* appears to be a superior form, as it is gradually expanding its range and forcing its sibling into less equitable habitats, and eventually extirpating it from an area entirely.

Evolutionary situations such as these can only be interpreted so completely because of the advanced nature of our knowledge and understanding of avian ecology and systematics. Many other examples undoubtedly can and do exist. However, where experimental breeding is impossible, it is only because of the secondary contact of the forms that we can deduce their differences, and thereby infer the evolutionary events that have produced them.

SYMPATRIC SPECIATION

It is generally believed that most cladogenetic evolution takes place by allopatric speciation: that is, a population or species is separated into two isolates and evolves separately therein. Here, the barrier is acting to reduce or even prevent gene flow. We have already discussed the way in which immigration from adjacent populations can markedly reduce the rates of inbreeding and random genetic drift at a neutral locus. It seems probable that only a very small number of migrants is sufficient to prevent even large populations from drifting and diverging away from each other.

We can show in a simple-minded way that migration can reduce the evolutionary divergence of populations subject to selection as well. Fig. 8.7 shows a series of populations that segregate for two alleles at a locus. In the populations that are numbered 1 to 6, the recessive phenotype is subjected to 100% mortality, and in 7 to 12, half of the dominant phenotype dies in each generation. The upper figure shows the changes in gene over six generations in the absence of gene flow. The populations on each side of the mid-point (or boundary between two ecological states) advance more or less uniformly away from their starting frequencies, with a marked discontinuity or step at the boundary. The lower diagram shows the results when there is 25% migration between adjacent populations. Again, there are clear differences between the two halves, but the step in the middle is appreciably less steep: gene flow is acting to reduce the divergence of the two sets of populations. No matter how long the model is left to evolve, there will always be migration of the 'wrong' gene across the boundary with a consequent production of the disadvantageous phenotypes and the maintenance of polymorphism.

The evolution of species in populations among which there is intermigration is known as sympatric speciation. It was believed for a long time that, because gene flow prevented (or at least reduced) diver-

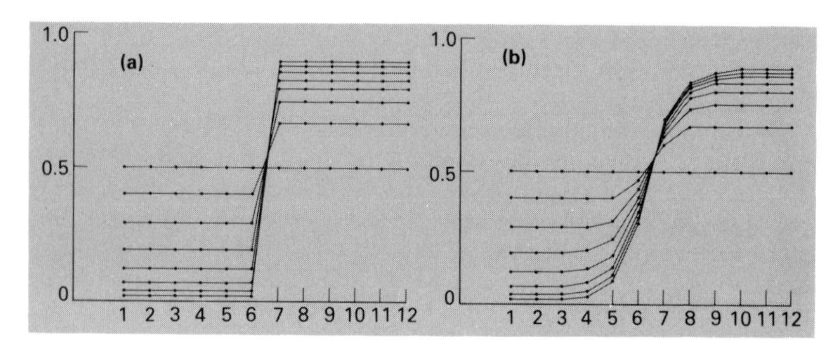

Fig. 8.7 The frequency of a recessive gene in a series of adjacent populations subject to strong selection over 7 generations from an initial frequency of 0.5. In populations 1 to 6, all of the recessive phenotypes die before reproduction; in the remainder, half of the dominant phenotypes die. (**a**) shows the results when there is no migration between adjacent populations, in (**b**) a greater number of the adults move into the adjacent population. Note the smoothing effect of gene flow upon the lower cline.

gence, speciation in its presence was well-nigh impossible, except of course in the case of polyploidy. So deep was this conviction that in 1963 Mayr wrote of sympatric speciation: 'One would think that it should no longer be necessary to devote much time to this topic, but past experience permits one to predict that the issue will be raised again and again at regular intervals.'[183] This was remarkably prescient of Mayr for, in the ten years subsequent to that statement, the subject of sympatric speciation was indeed raised in experimental, ecological and theoretical research. The impetus came from a series of studies in ecological genetics, of which we shall here consider two.

The first of these, which relates to the meadow brown butterfly (*Maniola jurtina*) must rank as one of the most curious and inexplicable phenomena in contemporary ecological genetics. This species occurs widely in western Europe and is particularly interesting because of the variation in the number of sub-terminal spots on the undersurface of the hind-wings. Males as a rule do not vary much from one population to the next, but females show a marked variation in some areas. The character is inherited, but heritability is low at cool temperatures such as the animals may experience in nature, although it rises appreciably at above-ambient conditions. It is interesting, and perhaps relevant, that heritability is significantly higher among the variable females than the more uniform males, suggesting there to be more genetic variation associated with the former.

Ford[102] reviews much of the published information relating to *M.*

jurtina, and shows that, over most of Britain and western Europe, there is virtually no variation in spot number between populations, and very little fluctuation with season. It seems as though a strong stabilizing selection might be holding the populations close to a genetic optimum. In some areas, however, this inter-population stability breaks down. For instance, in the Isles of Scilly off south-western Britain, populations on the large islands show a spot distribution different from those on the mainland (see Fig. 8.8). The smaller islands show differences not only from the large islands and the mainland, but from each other as well. Changes in spot number in a population do occasionally occur, usually when the environment has altered from some reason, and often when the populations are too large for random effects to be relevant. It is tempting to account for this on a selective model, but the principal agent or agents remain quite unknown.

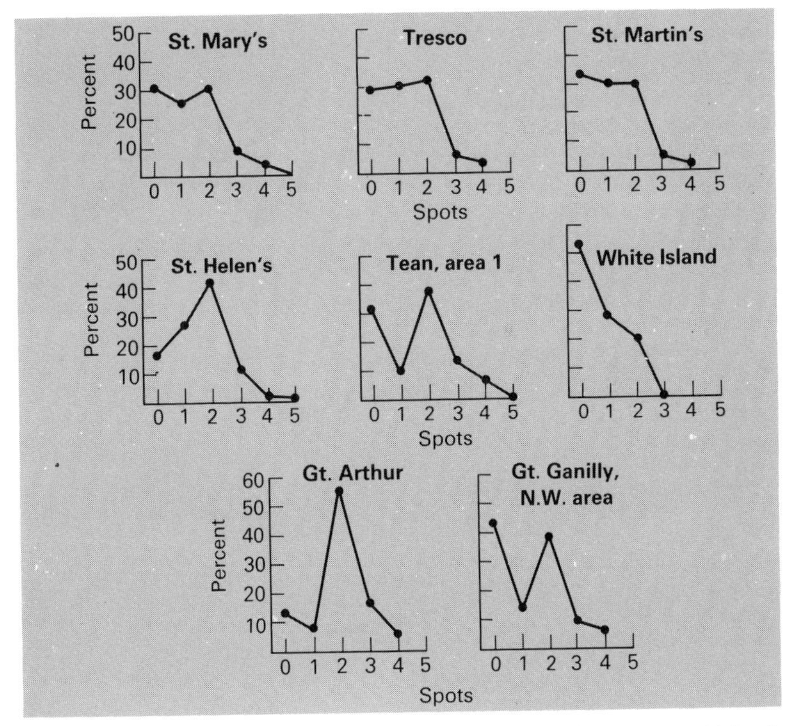

Fig. 8.8 Spot distribution of female *Maniola jurtina* on large and small islands in the Isles of Scilly. Note that the top three large populations are more uniform that the five smaller ones. From Ford.[102]

There is little else remarkable about the Scilly populations. Different spot distributions always occur in populations that are reproductively isolated from each other by ecological barriers such as the sea. On the mainland of south-western Britain, however, populations have been located that show marked differences in spot distribution in adjacent and continuous populations. Figure 8.9 shows a transect across the area in 1956 that demonstrates this rather well. The East and West Larrick populations are statistically very different, although they are separated only by a small hedge across which the

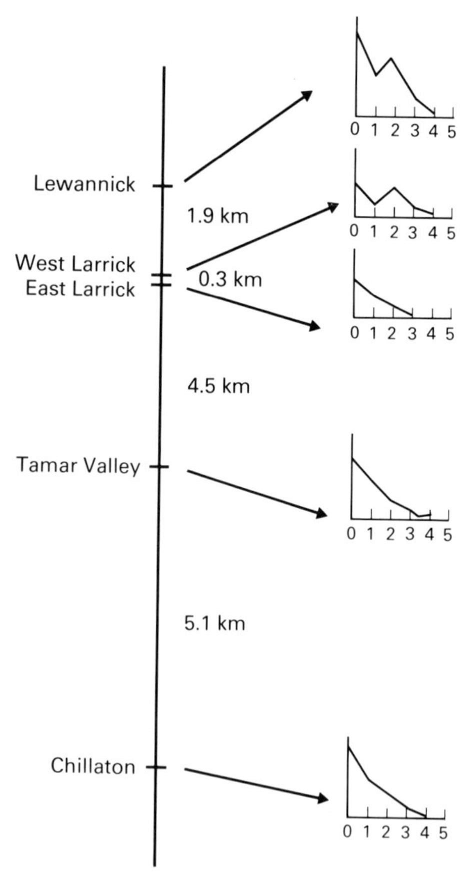

Fig. 8.9 The spot distribution of female *Maniola jurtina* in a series of populations from south Cornwall. The structure of the populations changes sharply at one point. Data from Ford.[102]

adults were flying freely. The following year, a similar pattern was found with bimodal populations in the west, and unimodal populations in the east. The boundary, however, had moved east by two to three miles. In subsequent years, it has moved to and fro by distances of up to forty miles between adjacent seasons.

Movement of *M. jurtina* is usually very limited, and changes in phenotype frequency over a range of so many miles from one season to the next can hardly be due to mass-migration. Nor can random processes be involved as possible causes, since the number of the animal's larvae must run to many thousands in most populations. Ford[102] suggests that these patterns are largely, if not wholly, the consequences of natural selection. Furthermore, since they occur in the presence of substantial gene flow (across the Larrick hedge, if nowhere else) they constitute an example of sympatric evolution, with populations showing divergence despite their proximity. The selective forces necessary to produce changes of the magnitude observed by Ford and his colleagues over distances of up to 40 miles must be quite enormous, and are difficult to envisage. It may be relevant to note that the normally uniform males sometimes show fluctuations in their spot distribution among populations in the boundary area. This might reflect a zone of genetic or developmental instability of some sort.

However, just because we cannot observe a selective force does not mean that one does not exist. Furthermore, Bradshaw and his colleagues have demonstrated repeatedly that, provided selective forces are sufficiently strong, such phenotypic divergence can occur in the presence of gene flow. We have already discussed some results of his studies of plants in extreme environments, and have seen the consequences of selection and gene flow on the heavy metal tolerance of plants living on and around mine spoil heaps in north Wales. This tolerance may have evolved very rapidly since some of the mines have only been worked for one or two hundred years. It is possible, however, that the soil contained above-average amounts of the metal prior to the site being worked. If this be the case, then there has been a longer period for selection to act.

Whatever the length of time involved, very interesting results have been reported by McNeilly and Antonovics[185] in *Agrostis tenuis* from Drws-y-Coed copper mine, and *Anthoxanthus odoratum* from Trelogan lead and zinc mine. These authors took a series of plants from points in a transect across the boundary of each mine and crossed them in the laboratory. They could find no evidence of incompatibility differences between the populations from mine and pasture soil. However, less seed was set in a cross between mine and pasture plants at Trelogan, when non-tolerant, pasture plants were used as the female parent. This may indicate some differentiation between the two populations.

A much more remarkable result is found when the flowering times of mine and pasture plants are compared along the same two transects. McNeilly and Antonovics[185] recorded the stage of flowering of plants using an arbitrary scale from 0 when the inflorescence was enclosed in the sheath, to 6 when the flowers were closed and the glumes were brown. Plotting the mean flowering stage on a particular date at Drws-y-Coed against position on the transect (Fig. 8.10) shows a marked discontinuity between sites 5 and 6, which is precisely the boundary between mine and pasture soil. It thus seems that *A. tenuis* plants growing on the contaminated soil of Drws-y-Coed mine have evolved an earlier flowering time than the adjacent pasture plants. That the difference is genetic is confirmed by the result holding in controlled greenhouse conditions.

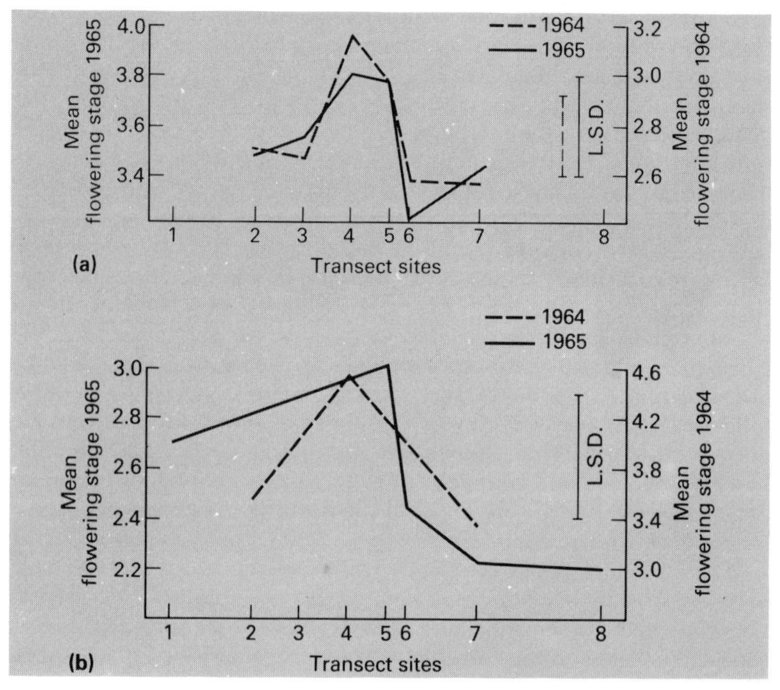

Fig. 8.10 Differences in flowering time in *Agrostis tenuis* growing at Drws-y-Coed copper mine in North Wales. The samples numbered 1 to 5 are from mine populations, and flower earlier than the rest. (**a**) Flowering time in field; (**b**) Flowering time in greenhouse. From McNeilly and Antonovics.[185]

Essentially the same results are found in *A. odoratum* from Trelogan (Fig. 8.11), with an anomalous point at site 8 that might be due to a slightly different sampling site. Nevertheless, it seems reasonably clear that plants growing on this mine also flower earlier in the season than those from pasture.

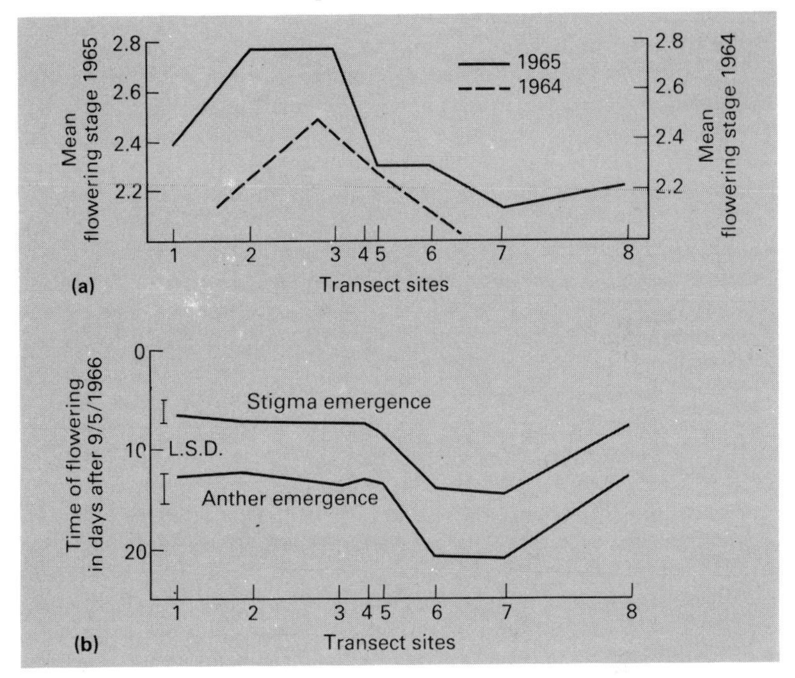

Fig. 8.11 As Fig. 8.10, but for *Anthoxanthum odoratum* on Trelogan lead and zinc mine. Again, plants from sites 1 to 4 come from contaminated soil, and flower earlier. (**a**) Flowering time in field; (**b**) Flowering time in garden. From McNeilly and Antonovics.[185]

McNeilly and Antonovics account for their findings as a combination of adaptations to local microclimate and towards a restriction of gene flow. They show that plants flower earlier in the warmer and drier sites, but that this alone is not sufficient to explain the differences. They go on to suggest that selection may be altering flowering time to reduce crossing between the mine and pasture populations, arguing as follows. There is evidence, reviewed by Jain and Bradshaw[130] that, if selection is sufficiently strong, divergence can occur. In both of the mines in the present study, there can be no doubt about the strength of selection. Seed that falls upon the mine soil must be tolerant, otherwise the seedling will die. Similarly, experimental analyses have suggested that tolerant grasses are less successful when competing with non-tolerant species in a closely pack sward on uncontaminated soil. Thus we can see that any mechanism to promote mine × mine or pasture × pasture crosses will be advantageous on the appropriate habitat, as they will produce offspring better adapted to the mine and pasture respectively.

The enhanced dryness of the mine soil might induce slightly earlier flowering in some individuals, and the plants that can respond to this genetically will fertilize and be fertilized by similarly tolerant neighbours. Conversely, those plants on the pasture that do not set seed until after the cessation of flowering by the mine population will not be fertilized by tolerant pollen. Following upon this response, the alleles for earlier flowering on the mine, and delayed flowering in the pasture close to the boundary will increase in frequency. Consequently, the populations will diverge in flowering time in response to selection imposed by the toxic effects of heavy metal contaminants in the soil.

Are we here seeing the beginnings of the evolution of two species where one stood before ? Time alone will tell, but certainly the work of McNeilly and Antonovics may be the first witnessing of the formation of breeding barriers between adjacent populations under conditions of heavy gene flow. Such a situation is known as parapatric speciation, to distinguish it from sympatric speciation where two species evolve in the same habitats, and allopatric where they evolve following isolation by some geographical or other barrier. McNeilly and Antonovics[185] also report that some plant species growing upon serpentine rock are reproductively isolated from their close relatives growing upon normal soil. Indeed, the slight incompatibility between the two populations of *A. odoratum* may be the beginning of a deeper genetical difference.

It can be argued that parapatric speciation only differs from allopatric in the existence of gene flow between the two populations of groups of populations. These are geographically separate, albeit with a common border, and each is well adapted to its own niche, and, most importantly, at a substantial disadvantage should it stray into the alien habitat. Indeed, at the extreme, if a migrant dies before breeding, the two populations are effectively allopatric despite gene flow between them. Evidently there is a 'grey area', where allopatry, sympatry and parapatry merge together under varying conditions of gene flow and selection.

However, in the absence of total selection against migrant individuals, groups of individuals that are capable of exchanging genes are not genetically or evolutionarily isolated from one another. This is one of the reasons why the disruptive selection experiments of Thoday and his colleagues (discussed in Chapter 5) are so interesting. Their experimental populations were subjected to massive gene flow, and yet they diverged. Maynard Smith[243] assessed this situation theoretically. He suggested that Thoday's results were inherently unstable because the 'hybrid' forms were less fit than the two 'parental' forms. However, he found that if these two parental forms were separately regulated in each niche, then stability could exist. The fitness of the two forms had to be different in each niche, and each had to be superior to the other in one. He also suggested that reproductive isolation could evolve following this, provided that habitat selection and preference occurred. He argued that this was likely to happen because those offspring that survived in a particular niche would be the ones that were best adapted to it. They were descended from parents that chose that niche, and the preference for it should be reinforced.

Examples of this in nature are hard to come by. Ford *et al.*[103] suggested that sympatric speciation along these lines might be occurring among Darwin's finches of the species *Geospiza fortis* on Santa Cruz in the Galapagos Islands. Although there is considerable variation between individuals, the species exists as two forms, large-billed and small-billed. This difference is not related to age or sex, but is probably important in feeding, for the analysis of stomach contents of *Geospiza* finches suggests that the size of food taken depends upon the size and conformation of the bill, although again there is some individual variation. Kear[143] has shown that individual species of European finches vary in their handling ability of different seeds. There is an optimum food size and hardness for a particular species. Ford *et al.*[103] suggest that the existence of 'small' and 'medium' sizes of seeds might constitute two different niches, and that the two groups of *G. fortis* are differentially adapted to them. They refer to behavioural studies by Lack[165] that showed bill size to be an important component of species recognition in *Geospiza* and cite these as supporting evidence that assortative mating might act to reduce gene flow between the two groups.

A detailed ecological and behavioural study by Grant *et al.*[113] showed that flocks of *G. fortis* differed in their bill size. This would increase assortative mating if pair-formation took place within these flocks. They considered that this supported the hypothesis of Ford *et al.*, but added that it could also be due to allopatric divergence. If the two size classes had originated on separate islands, and subsequently came into secondary contact, a similar pattern of differential flock composition might result. Partial genetic isolation between the small and large *G. fortis*, having arisen during a period of allopatry, would then be maintained in the present sympatric situation by these behavioural differences. Whether this maintenance is permanent or temporary is, of course, not known. The whole matter remains one for conjecture, but despite the lack of hard field evidence, the phenomenon of sympatric speciation remains at least a theoretical possibility in certain special situations.

References

1. ALLEN, J. A. and CLARKE, B. (1968). Evidence for apostatic selection by wild passerines. *Nature, Lond.*, **220**, 501–2.
2. ALLISON, A. C. (1954). Protection afforded by the sickle-cell trait against subtertian malarial infection. *Br. Med. J.*, **1**, 290–4.
3. ALLISON, A. C. (1955). Aspects of polymorphism in man. *Cold Spr. Harb. Symp. Quant. Biol.*, **20**, 239–55.
4. ALLISON, A. C. (1964). Polymorphism and natural selection in human populations. *Cold Spr. Harb. Symp. Quant. Biol.*, **29**, 137–49.
5. ANXOLABEHERE, D. and PERIQUET, G. (1972). Variation de la valeur selective de l'heterozygote en fonction de frequences alleliques chez *Drosophila melanogaster. C.R. Acad. Sci, Paris,* **275**, 2755–7.
6. ARNOLD, R. W. (1966). *Factors affecting the gene-frequencies of British and continental populations of* Cepaea. Doctoral Thesis, Bodleian Library, Oxford.
7. ASTON, J. L. and BRADSHAW, A. D. (1966). Evolution in closely adjacent plant populations. II. *Agrostis stolonifera* in maritime habitats. *Heredity*, **21**, 649–64.
8. AVISE, J. C. and SELANDER, R. K. (1972). Evolutionary genetics of cave-dwelling fishes of the genus *Astyanax. Evolution*, **26**, 1–18.
9. AYALA, F. J. (1975). Genetic differentiation during the speciation process. In: *Evolutionary Biology*, Vol. **8**. Th. Dobzhansky, M. K. Hecht and W. C. Steere (eds.), Plenum Press, New York.
10. AYALA, F. J., TRACEY, M. L., BARR, L. G., MCDONALD, J. F. and PEREZ-SALAS, S. (1974). Genetic variation in natural populations of five *Drosophila* species and the hypothesis of the selective neutrality of protein polymorphism. *Genetics*, **77**, 343–84.
11. BASTOCK, M. and MANNING, A. (1955). The courtship of *Drosophila melanogaster. Behaviour*, **8**, 85–111.
12. BATES, H. W. (1863). *The naturalist on the river Amazon*, Vol. **1**, John Murray, London.
13. BERNSTEIN, S. C., THROCKMORTON, L. H. and HUBBY, J. L. (1973). Still more genetic variability in natural populations. *Proc. Nat. Acad. Sci*, **70**, 3928–31.

14. BISHOP, J. A. and COOK, L. M. (1975). Moths, melanism and clean air. *Sci. Amer.*, **232**, 90–9.
15. BRADSHAW, A. D. (1960). Population differentiation in *Agrostis tenuis* Sibth. III. Populations in varied environments. *New Phytol.*, **59**, 92–103.
16. BRAUNITZER, G., HILSE, K., RUDOLFF, V. and HILSCHMANN, N. (1964). The haemoglobins. *Adv. Protein Chemistry*, **19**, 1–71.
17. BRNCIC, D. (1954). Heterosis and the integration of the genotype in geographic populations of *D. pseudoobscura*. *Genetics*, **39**, 77–88.
18. BRNCIC, D. and KOREF-SANTIBAENZ, S. (1964). Mating activity of homo- and heterokaryotypes in *Drosophila pavani*. *Genetics*, **49**, 585–91.
19. BROWER, J. V. Z. (1958). Experimental studies of mimicry in some North American butterflies. Part II. *Battus philenor* and *Papilio troilus, P. polyxenes* and *P. glaucus*. *Evolution*, **12**, 123–36.
20. BROWER, J. V. Z. (1960). Experimental studies of mimicry. IV. The reactions of starlings to different proportions of models and mimics. *Amer. Nat.*, **94**, 271–82.
21. BROWER, L. P., BROWER, J. VAN Z. and COLLINS, C. T. (1963). Relative palatability and Müllerian mimicry among neotropical butterflies of the subfamily Heliconiinae. *Zoologica, N.Y.*, **48**, 65–84.
22. BROWER, L. P., COOK, L. M. and CROZE, H. J. (1967). Predator response to artificial Batesian mimics released in a Neotropical environment. *Evolution*, **21**, 11–23.
23. BUMPUS, H. C. (1898a). The variations and mutations of the introduced sparrow. *Biol. Lectures, Marine Biol. Lab., Woods Hole*, 1–15.
24. BUMPUS, H. C. (1898b). The elimination of the unfit as illustrated by the introduced sparrow. *Biol. Lectures, Marine Biol. Lab., Woods Hole*, 209–26.
25. BURI, P. (1956). Gene frequency in small populations of mutant *Drosophila*. *Evolution*, **10**, 367–402.
26. CAIN, A. J. (1954). *Animal Species and their Evolution*. Hutchinson, London.
27. CAIN, A. J. and CURREY, J. D. (1963a). Differences in interaction between selective forces acting in the wild on certain pleiotropic genes of *Cepaea*. *Nature, Lond.*, **197**, 411–12.
28. CAIN, A. J. and CURREY, J. D. (1963b). Area effects in *Cepaea*. *Phil. Trans. Roy. Soc. Lond. B*, **246**, 1–81.
29. CAIN, A. J. and CURREY, J. D. (1963c). The causes of area effects. *Heredity*, **18**, 459–71.
30. CAIN, A. J., KING, J. M. B. and SHEPPARD, P. M. (1960). New data on the genetics of polymorphism in the snail *Cepaea nemoralis* L. *Genetics*, **45**, 393–411.
31. CAIN, A. J. and SHEPPARD, P. M. (1950). Selection in the polymorphic land snail *Capaea nemoralis* (L.). *Heredity*, **4**, 275–94.
32. CAIN, A. J. and SHEPPARD, P. M. (1954). Natural selection in *Cepaea*. *Genetics*, **39**, 89–116.
33. CAIN, A. J., SHEPPARD, P. M. and KING, J. M. B. (1968). Studies on *Cepaea*. I. The genetics of some morphs and varieties of *Cepaea nemoralis*. *Phil. Trans. Roy. Soc. Lond. B*, **253**, 383–96.
34. CANNON, H. G. (1959). *Lamarck and Modern Genetics*. Manchester University Press, Manchester.
35. CEPPELLINI, R. (1955). Discussion of 'Aspects of polymorphism in man'

by A. C. Allison. *Cold Spr. Harb. Symp. Quant. Biol.*, **20**, 252–5.

36. CHARLESWORTH, B. (1971). Selection in density-regulated populations. *Ecology*, **52**, 469–75.
37. CHARLESWORTH, B. and CHARLESWORTH, D. (1973). A study of linkage disequilibrium in populations of *Drosophila melanogaster*. *Genetics*, **73**, 351–9.
38. CLARKE, B. (1960). Divergent effects of natural selection on two closely-related polymorphic snails. *Heredity*, **14**, 423–43.
39. CLARKE, B. (1962). Balanced polymorphism and the diversity of sympatric species. *Systematics Assoc. Publ.*, **4**, 47–70.
40. CLARKE, B. C. (1968). Balanced polymorphism and regional differentiation in landsnails. In *Evolution and Environment*. ed. DRAKE, E. T., Yale University Press, London.
41. CLARKE, B. (1969). The evidence for apostatic selection. *Heredity*, **24**, 347–52.
42. CLARKE, B. (1970a). Darwinian evolution of proteins. *Science, N.Y.*, **168**, 1009–11.
43. CLARKE, B. (1970b). Selective constraints on amino-acid substitutions during the evolution of proteins. *Nature, Lond.*, **228**, 159–60.
44. CLARKE, B. (1972). Density-dependent selection. *Amer. Nat.*, **106**, 1–13.
45. CLARKE, B. (1973). Mutation and population size. *Heredity*, **31**, 367–79.
46. CLARKE, B. (1975). The contribution of ecological genetics to evolutionary theory: detecting the direct effects of natural selection on particular polymorphic loci. *Genetics*, **79**, 101–13.
47. CLARKE, B. C. and MURRAY, J. J. (1971). Polymorphism in a Polynesian land snail, *Partula suturalis vexillum*. In *Ecological Genetics and Evolution*, ed. CREED, E. R., Blackwells, Oxford.
48. CLARKE, B. and O'DONALD, P. (1964). Frequency-dependent selection. *Heredity*, **19**, 201–6.
49. CLARKE, C. A. (1964). *Genetics for the Clinician*. 2nd edn. Blackwell Scientific Publications, Oxford.
50. CLARKE, C. A. and SHEPPARD, P. M. (1960a). The Genetics of *Papilio dardanus*, Brown. II. Races Dardanus, Polytrophus, Meseres and Tibullus. *Genetics*, **45**, 439–57.
51. CLARKE, C. A. and SHEPPARD, P. M. (1960b). The evolution of dominance under disruptive selection. *Heredity*, **14**, 73–87.
52. CLARKE, C. A. and SHEPPARD, P. M. (1960c). The evolution of mimicry in the butterfly *Papilio dardanus*. *Heredity*, **14**, 163–73.
53. CLARKE, C. A. and SHEPPARD, P. M. (1960d). Supergenes and mimicry. *Heredity*, **14**, 175–85.
54. CLARKE, C. A., SHEPPARD, P. M. and THORNTON, I. W. B. (1968). The genetics of the mimetic butterfly *Papilio memnon*. *Phil. Trans. Roy. Soc. Lond. B*, **254**, 37–89.
55. COCKERHAM, C. C., BURROWS, P. M., YOUNG, S. S. and PROUT, T. (1972). Frequency-dependent selection in randomly mating populations. *Amer. Nat.*, **106**, 493–515.
56. COOK, L. M., ASKEW, R. R. and BISHOP, J. A. (1970). Increasing frequency of the typical form of the Peppered Moth in Manchester. *Nature, Lond.*, **227**, 1155.
57. COOK, L. M., BROWER, L. P. and ALCOCK, J. (1969). An attempt to verify mimetic advantage in a neotropical environment. *Evolution*, **23**, 339–45.

58. CORRENS, C. (1900). G. Mendels Regel über das Verhalten der Nachkommenschaft der Rassenbastarde. *Ber. dt. bot. Ges.*, **18**, 158–68.
59. CROSBY, J. L. (1940). High proportions of homostyle plants in populations of *Primula vulgaris*. *Nature, Lond.*, **145**, 672–3.
60. CROW, J. F. (1961). Population genetics. *Am. J. Human Genetics*, **13**, 137–50.
61. CROW, J. F. and KIMURA, M. (1970). *An Introduction to Population Genetics Theory*. Harper and Row, New York.
62. CUELLAR, H. S. (1971). Levels of genetics compatibility of *Rana areolata* with southwestern members of the *Rana pipiens* complex (Anura: Ranidae). *Evolution*, **25**, 399–409.
63. CURREY, J. D. and CAIN, A. J. (1968). Studies on *Cepaea*. 4. Climate and selection of banding morphs in *Cepaea* from the climatic optimum to the present day. *Phil. Trans. Roy. Soc. Lond. B*, **253**, 483–98.
64. DARWIN, C. (1859). *On the Origin of Species by Means of Natural Selection*. John Murray, London.
65. DARWIN, C. (1889). *A Naturalist's Voyage*. John Murray, London.
66. DAVIDSON, J. (1938). On the growth of the sheep population in Tasmania. *Trans. R. Soc. South Australia*, **62**, 342–6.
67. DAWOOD, M. M. and STRICKBERGER, M. W. (1969). The effect of larval interaction on viability in *Drosophila melanogaster*. III. Effects of biotic residues. *Genetics*, **63**, 213–20.
68. DAY, T. H. (1972). A comparison of the variability of muscle, hair, lens and plasma proteins. *Experientia*, **28**, 1154–5.
69. DAY, T. H. and CLAYTON, R. M. (1973). Intraspecific Variation in the Lens Proteins. *Biochem. Genet.*, **8**, 187–203.
70. DAY, T. H., HILLIER, P. C. and CLARKE, B. C. (1974). Properties of genetically polymorphic isozymes of alcohol dehydrogenase in *Drosophila melanogaster*. *Biochem. Genet.*, **11**, 141–53.
71. DAYHOFF, M. O. (1972). *Atlas of Protein Sequence and Structure*. Volume 5. National Biomedical Research Council, Washington, D.C.
72. DAYHOFF, M. O., PARK, C. M. and MCLAUGHLIN, P. J. (1972). Building a phylogenetic tree: cytochrome C. In *Atlas of Protein Sequence and Structure* Volume 5. ed. DAYHOFF, M. O., National Biomedical Research Council, Washington, D.C.
73. DE VRIES, H. (1900). Das Spaltungsgesetz der Bastarde. *Ber. dt. bot. Ges.*, **18**, 83–90.
74. DIVER, C. (1940). The problem of closely related species living together in the same area. In *The New Systematics*. ed. HUXLEY, JULIAN, Oxford.
75. DOBZHANSKY, TH. (1950). Mendelian populations and their evolution. *Amer. Nat.*, **84**, 401–18.
76. DOBZHANSKY, TH. (1951). *Genetics and the Origin of Species*. Rev. 3rd Edition. Columbia University Press, New York.
77. DOBZHANSKY, TH. (1952). The nature and origin of heterosis. In *Heterosis*, ed. GOWEN, J. W., Iowa State College Press, Iowa.
78. DOBZHANSKY, TH. (1970). *Genetics of the Evolutionary Process*. Columbia University Press, New York.
79. DOBZHANSKY, TH. and PAVLOVSKY, O. (1953). Indeterminate outcome of certain experiments on *Drosophila* populations. *Evolution*, **7**, 198–210.
80. DOBZHANSKY, TH. and WRIGHT, S. (1943). Genetics of natural popu-

lations. X. Dispersion rates in *Drosophila pseudoobscura*. *Genetics*, **28**, 304–40.
81. EHRMAN, L. (1966). Mating success and genotype frequency in *Drosophila*. *Anim. Behav.*, **14**, 332–9.
82. EHRMAN, L. (1967). Further studies on genotype frequency and mating success in *Drosophila*. *Amer. Nat.*, **101**, 415–24.
83. EHRMAN, L. (1968). Frequency dependence of mating success in *Drosophila pseudoobscura*. *Genet. Res.*, **11**, 135–40.
84. EHRMAN, L. (1969). The sensory basis of mate selection in *Drosophila*. *Evolution*, **23**, 59–64.
85. EINARSEN, A. S. (1942). Specific results from ring-necked pheasant studies in the Pacific north west. *Trans. North Amer. Wildl. Conf.*, **7**, 130–45.
86. EINARSEN, A. S. (1945). Some factors affecting ring-necked pheasant population density. *Murrelet*, **26**, 39–44.
87. EMERSON, S. (1939). A preliminary survey of the *Oenothera organensis* population. *Genetics*, **24**, 524–37.
88. EPLING, C. and DOBZHANSKY, TH. (1942). Genetics of natural populations. VI. Microgeographic races in *Linanthus parryae*. *Genetics*, **27**, 317–32.
89. EPLING, C., LEWIS, H. and BALL, F. M. (1960). The breeding group and seed storage: a study in population dynamics. *Evolution*, **14**, 238–55.
90. ERNST, A. (1936). Weitere Untersuchungen zur Phänanalyse zum Fertilitätsproblem und zur Genetik Heterostyler Primeln. II. *Primula hortensis*. *Arch. Klaus-Stift. Vererb-Forsch.*, **11**, 1–280.
91. EWENS, W. J. (1964). On the problem of self-sterility alleles. *Genetics*, **50**, 1433–8.
92. EWENS, W. H. (1972). The sampling theory of selectively neutral genes. *Theor. Pop. Biol.*, **3**, 87–112.
93. FISHER, R. A. (1927). On some objections to mimicry theory; statistical and genetic. *Trans. R. Ent. Soc. Lond.*, **75**, 269–78.
94. FISHER, R. A. (1928a). The possible modification of the response of the wild type to recurrent mutations. *Amer. Nat.*, **62**, 115–26.
95. FISHER, R. A. (1928b). Two further notes on the origin of dominance. *Amer. Nat.*, **62**, 571–4.
96. FISHER, R. A. (1930). *The Genetical Theory of Natural Selection*. Clarendon Press, Oxford.
97. FISHER, R. A. and FORD, E. B. (1947). The spread of a gene in natural conditions in a colony of the moth *Panaxia dominula* L. *Heredity*, **1**, 143–74.
98. FITCH, W. M. and MARGOLIASH, E. (1967). A method for estimating the number of invariant amino acid coding positions in a gene using cytochrome C as a model case. *Biochem. Genet.*, **1**, 65–71.
99. FITCH, W. M. and MARKOWITZ, E. (1970). An improved method for determining codon variability in a gene, and its application to the rate of fixation of mutations in evolution. *Biochem. Genet.*, **4**, 579–93.
100. FORD, E. B. (1940). Polymorphism and taxonomy. In *The New Systematics*, pp. 493–513, ed. HUXLEY, JULIAN, Clarendon Press, Oxford.
101. FORD, E. B. (1940). Genetic research in the Lepidoptera. *Ann. Eugen.*, **10**, 227–52.
102. FORD, E. B. (1975). *Ecological Genetics*. 4th edn. Methuen, London.

103. FORD, H. A., PARKIN, D. T. and EWING, A. W. (1973). Divergence and evolution in Darwin's Finches. *Biol. J. Linn. Soc.*, **5**, 289–95.
104. FRANKLIN, L. and LEWONTIN, R. C. (1970). Is the gene the unit of selection? *Genetics*, **65**, 707–34.
105. FRASER, J. F. D. and ROTHSCHILD, THE HON. M. (1960a). Defence mechanisms in warningly-coloured moths and other insects. *XI Int. Kongr. Fr. Entom. Wien, BIII, (Symposium 4)*, 249–56.
106. FRASER, J. F. D. and ROTHSCHILD, THE HON. M. (1960b). Defence mechanisms in warningly-coloured moths and other insects. *Proc. 11th Int. Cong. Entom.*, **3**, 249–56.
107. GEIGER, R. (1959). *The Climate Near the Ground.* Harvard University Press, Cambridge, Mass.
108. GIBSON, J. B. (1970). Enzyme flexibility in *Drosophila melanogaster. Nature, Lond.*, **227**, 959.
109. GILLESPIE, J. H. and LANGLEY, C. H. (1974). A general model to account for enzyme variation in natural populations. *Genetics*, **76**, 837–84.
110. GOLDSCHMIDT, R. B. (1940). *The Material Basis of Evolution.* Yale University Press, New Haven.
111. GOODHART, C. B. (1963). 'Area effects' and non-adaptive variation between populations of *Cepaea* (Mollusca). *Heredity*, **18**, 459–71.
112. GORDON, C. (1935). An experiment on a released population of *D. melanogaster. Amer. Nat.*, **69**, 381.
113. GRANT, P. R., GRANT, B. R., SMITH, J. N. M., ABBOTT, I. J. and ABBOTT, L. K. (1976). Darwin's finches: Population variation and natural selection. *Proc. Nat. Acad. Sci. U.S.A.*, **73**, 257–61.
114. GREAVES, J. H. (1972). Resistance to anticoagulants in rats. *Pestic. Sci*, **2**, 276–9.
115. GREGORY, R. P. G. and BRADSHAW, A. D. (1965). Heavy metal tolerance in populations of *Agrostis tenuis* and other grasses. *New Phytol.*, **64**, 131–43.
116. HALDANE, J. B. S. (1924). A mathematical theory of natural and artificial selection. *Trans. Camb. Phil. Soc.*, **23**, 19–40.
117. HALDANE, J. B. S. (1927a). A mathematical theory of natural and artificial selection. Part IV. *Proc. Camb. Phil. Soc.*, **23**, 607–15.
118. HALDANE, J. B. S. (1927b). A mathematical theory of natural and artificial selection. Part V. Selection and mutation. *Proc. Camb. Phil. Soc.*, **23**, 838–44.
119. HALDANE, J. B. S. (1939). The theory of the evolution of dominance. *J. Genet.*, **37**, 365–74.
120. HALDANE, J. B. S. (1957). The cost of natural selection. *J. Genet.*, **57**, 351–60.
121. HALDANE, J. B. S. and JAYAKAR, S. D. (1963). Polymorphism due to selection depending on the composition of a population. *J. Genet.*, **58**, 318–23.
122. HARDING, J. (1970). Genetics of *Lupinus.* II. The selective disadvantage of the pink flower colour mutant in *Lupinus nanus. Evolution*, **24**, 120–27.
123. HARDING, J., ALLARD, R. W. and SMELTZER, D. G. (1966). Population studies in predominantly self-pollinated species. 9. Frequency-dependent selection in *Phaseolus lunatus. Proc. Nat. Acad. Sci. U.S.A.*, **56**, 99–104.
124. HARRIS, H. (1966). Enzyme polymorphisms in man. *Proc. Roy. Soc. Lond.*

B, **164**, 298–310.

125. HEBERT, P. D. N., WARD, R. D. and GIBSON, J. B. (1972). Natural selection for enzyme variants among parthenogenetic *Daphnia magna*. *Genet. Res.*, **19**, 173–6.
126. HUBBY, J. L. and LEWONTIN, R. C. (1966). A molecular approach to the study of genic heterozygosity in natural populations. I. The number of alleles at different loci in *Drosophila pseudoobscura*. *Genetics*, **54**, 577–94.
127. HUBBY, J. L. and THROCKMORTON, L. H. (1968). Protein differences in *Drosophila*. IV. A study of sibling species. *Amer. Nat.*, **102**, 193–205.
128. IMBRIE, J. (1957). The species problem with fossil animals. In *The Species Problem*, ed. MAYR, E., *Amer. Assoc. Advanc. Sci. Publ.*, **50**, 125–52.
129. INGRAM, V. M. (1963). *The Hemoglobins in Genetics and Evolution*. Columbia University Press, New York.
130. JAIN, S. K. and BRADSHAW, A. D. (1966). Evolutionary divergence in adjacent plant populations. I. The evidence and its theoretical analysis. *Heredity*, **21**, 407–41.
131. JOHANSSON, I. and RENDEL, J. (1968). *Genetics and Animal Breeding*. Oliver and Boyd, Edinburgh.
132. JOHNSON, F. M. and POWELL, A. (1974). The alcohol dehydrogenases of *Drosophila melanogaster*: frequency changes associated with heat and cold shock. *Proc. Nat. Acad. Sci. U.S.A.*, **71**, 1783–4.
133. JOHNSON, G. B. (1973). Importance of substrate variability to enzyme polymorphisms. *Nature, New Biol.*, **243**, 151–3.
134. JOHNSON, G. B. and FELDMAN, M. W. (1973). On the hypothesis that polymorphic enzyme alleles are selectively neutral. I. The evenness of allele frequency distribution. *Theoret. Pop. Biol.*, **4**, 209–21.
135. JOHNSTON, R. F., NILES, D. M. and ROHWER, S. A. (1972). Herman Bumpus and natural selection in the House Sparrow *Passer domesticus*. *Evolution*, **26**, 20–31.
136. JOHNSTON, R. F. and SELANDER, R. K. (1971). Evolution in the House Sparrow. II. Adaptive differentiation in North American populations. *Evolution*, **25**, 1–28.
137. JONES, J. S. and YAMAZAKI, T. (1974). Genetic background and the fitness of allozymes. *Genetics*, **78**, 1185–9.
138. JOWETT, D. (1964). Population studies on lead-tolerant *Agrostis tenuis*. *Evolution*, **18**, 70–81.
139. JUKES, T. H. (1972). Comparison of polypeptide sequences. In *Proc. 6th Berkeley Symp. on Math. Stats. and Prob. V. Darwinian, Neo-Darwinian and Non-Darwinian Evolution*, eds. LE CAM, L. M., *et al.* pp. 101–27, University California Press, Berkeley.
140. KAMMERER, P. (1919). Vererbung erzwungener Formveränderungen. I. Mitteilung: Brunstschwiele der Alytes – Männchen aus "Wasseretern" (Zugleich: Vererbung erzwungener Fortpflanzungsanpassungen. V. Mitteilung). *Arch. fur Entwicks.*, **45**, 323–70.
141. KAMMERER, P. (1923). Breeding experiments on the inheritance of acquired characters. *Nature, Lond.*, **111**, 637–40.
142. KARN, M. N. and PENROSE, L. S. (1951). Birth-weight and gestation time in relation to maternal age, parity and infant survival. *Ann. Eugen.*, **16**, 147–64.

143. KEAR, J. (1962). Food selection in finches, with special reference to interspecific differences. *Proc. zool. Soc. Lond.*, **138**, 163–205.
144. KERR, W. E. and WRIGHT, S. (1954a). Experimental studies of the distribution of gene frequencies in very small populations of *Drosophila melanogaster*. I. Forked. *Evolution*, **8**, 172–7.
145. KERR, W. E. and WRIGHT, S. (1954b). Experimental studies of the distribution of gene frequencies in very small populations of *Drosophila melanogaster*. II. Bar. *Evolution*, **8**, 225–40.
146. KETTLEWELL, H. B. D. (1958). Industrial melanism in the Lepidoptera and its contribution to our understanding of evolution. *Proc. 10th Int. Congr. Ent.*, **2**, 831–41.
147. KETTLEWELL, H. B. D. (1965). Insect survival and selection for pattern. *Science, N.Y.*, **148**, 1290–6.
148. KETTLEWELL, H. B. D. (1973). *The Evolution of Melanism*. Clarendon Press, Oxford.
149. KIMURA, M. (1957). Some problems of stochastic processes in genetics. *Ann. Math. Stat.*, **28**, 882–901.
150. KIMURA, M. (1968). Evolutionary rate at the molecular level. *Nature, Lond.*, **217**, 624–6.
151. KIMURA, M. (1969). The rate of molecular evolution considered from the standpoint of population genetics. *Proc. Nat. Acad. Sci. U.S.A.*, **63**, 1181–8.
152. KIMURA, M. (1970). The length of time required for a selectively neutral mutant to reach fixation through random frequency drift in a finite population. *Genet. Res.*, **15**, 131–3.
153. KIMURA, M. and CROW, J. F. (1964). The number of alleles that can be maintained in a finite population. *Genetics*, **49**, 725–38.
154. KIMURA, M. and OHTA, T. (1971). *Theoretical Aspects of Population Genetics*. Princeton University Press, Princeton.
155. KING, J. L. (1967). Continuously distributed factors affecting fitness. *Genetics*, **55**, 483–92.
156. KING, J. L. (1972). The role of mutation in evolution. *Proc. 6th Berkeley Symp. on Math. Stats. and Prob. V. Darwinian, Neo-Darwinian and Non-Darwinian Evolution*, eds. LE CAM, L. M., *et al.*, California University Press, Berkeley, pp. 69–100.
157. KING, J. L. (1974). Isoallele frequencies in very large populations. *Genetics*, **76**, 607–13.
158. KING, J. L. and JUKES, T. H. (1969). Non-Darwinian evolution: random fixation of selectively neutral mutations. *Science, N.Y.*, **164**, 788–98.
159. KLITZ, W. (1973). Empirical population genetics of the North American House Sparrow. *Ornithological Monographs*, **14**, 39–48.
160. KOESTLER, A. (1971). *The Case of the Midwife Toad*. Hutchinson, London.
161. KOJIMA, K., GILLESPIE, J. and TOBARI, Y. N. (1970). A profile of *Drosophila* species enzymes assayed by electrophoresis. I. Number of alleles, heterozygosities and linkage disequilibrium in glucose-metabolising systems and some other enzymes. *Biochem. Genet.*, **4**, 627–37.
162. KOJIMA, K. and HUANG, S. L. (1972). Effects of population density on the frequency-dependent selection in the Esterase-6 locus of *Drosophila melanogaster*. *Evolution*, **26**, 313–21.
163. KOJIMA, K. and TOBARI, Y. N. (1969). The pattern of viability changes associated with genotype frequency at the alcohol dehydrogenase

locus in a population of *Drosophila melanogaster*. *Genetics*, **61**, 201–9.
164. KOJIMA, K. and YARBROUGH, K. M. (1967). Frequency-dependent selection at the Esterase-6 locus in *Drosophila melanogaster*. *Proc. Nat. Acad. Sci. U.S.A.*, **57**, 645–9.
165. LACK, D. (1947). *Darwin's Finches*. Cambridge University Press, Cambridge.
166. LACK, D. (1954). *The Natural Regulation of Animal Numbers*. Oxford University Press, London.
167. LACK, D. (1971). *Ecological Isolation in Birds*. Blackwell, Oxford.
168. LAMOTTE, M. (1951). Recherches sur la structure génétique des populations naturelles de *Cepaea nemoralis* (L.) *Bull. Biol. Suppl.*, **35**, 1–239.
169. LERNER, I. M. (1954). *Genetic Homeostasis*. John Wiley, New York.
170. LEVIN, D. A. (1972). Low frequency disadvantage in the exploitation of pollinators by corolla variants in *Phlox*. *Amer. Nat.*, **106**, 453–60.
171. LEVIN, D. A. and KERSTER, H. W. (1973). Assortative pollination for stature in *Lythrum salicania*. *Evolution*, **27**, 144–52.
172. LEVITAN, M. and MONTAGU, A. (1971). *Textbook of Human Genetics*. Oxford University Press, Oxford.
173. LEWIS, D. (1948). Structure of the incompatibility gene. 1. Spontaneous mutation rate. *Heredity*, **2**, 219–36.
174. LEWIS, K. R. and JOHN, B. (1970). *The Organisation of Heredity*. Edward Arnold, London.
175. LEWONTIN, R. C. (1972). Testing the Theory of Natural Selection. Review of *Ecological Genetics and Evolution*, ed. CREED, R., Blackwell, Oxford. *Nature, Lond.*, **236**, 181–2.
176. LEWONTIN, R. C. and HUBBY, J. L. (1966). A molecular approach to the study of genic heterozygosity in natural populations. II. Amount of variation and degree of heterozygosity in natural populations of *Drosophila pseudoobscura*. *Genetics*, **54**, 595–609.
177. LI, C. C. (1955). *Poulation Genetics*. University of Chicago Press, Chicago.
178. LOWTHER, J. K. (1961). Polymorphism in the white-throated sparrow, *Zonotrichia albicollis*. *Canad. J. Zool.*, **39**, 281–92.
179. MACFARQUHAR, A. M. and ROBERTSON, F. W. (1963). The lack of evidence for coadaptation in crosses between geographical races of *Drosophila subobscura* Coll. *Genet. Res.*, **4**, 104–31.
180. MANLY, B. F. J., MILLER, P. and COOK, L. M. (1972). Analysis of a selective predation experiment. *Amer. Nat.*, **106**, 719–36.
181. MATHER, K. (1955). Polymorphism as an outcome of disruptive selection. *Evolution*, **9**, 52–61.
182. MAYR, E. (1942). *Systematics and the Origin of Species*. Columbia University Press, New York.
183. MAYR, E. (1963). *Animal Species and Evolution*. Harvard University Press, Cambridge, Mass.
184. MCNEILLY, T. S. (1968). Evolution in closely adjacent plant populations. III. *Agrostis tenuis* on a small copper mine. *Heredity*, **23**, 99–108.
185. MCNEILLY, T. and ANTONOVICS, J. (1968). Evolution in closely adjacent plant populations. IV. Barriers to gene flow. *Heredity*, **23**, 205–18.
186. MENDEL, G. (1866). Versuche über Pflanzen hybriden. *Verh. naturf. Ver. Brünn*, **4**, 3–44.
187. MILKMAN, R. D. (1967). Heterosis as a major cause of heterozygosity in nature. *Genetics*, **55**, 493–5.

188. MILLER, M. J., NEEL, J. V. and LIVINGSTONE, F. B. (1956). Distribution of parasites in the red cells of sickle-cell trait carriers infected with *Plasmodium falciparum*. *Trans. Roy. Soc. Trop. Med. Hyg.*, **50**, 294–6.
189. MITTON, J. B. and KOEHN, R. K. (1973). Population genetics of marine pelecypods. III. Epistasis between functionally related isoenzymes of *Mytilus edulis*. *Genetics*, **73**, 487–96.
190. MOORE, J. A. (1946). Hybridisation between *Rana palustris* and different geographic forms of *Rana pipiens*. *Proc. Nat. Acad. Sci. U.S.A.*, **32**, 209–12.
191. MOORE, J. A. (1947). Hybridisation between *Rana pipiens* from Vermont and eastern Mexico. *Proc. Nat. Acad. Sci, U.S.A.*, **33**, 72–5.
192. MOORE, J. A. (1949). Geographic variation of adaptive characters in *Rana pipiens* Schreber. *Evolution*, **3**, 1–24.
193. MOORE, J. A. (1950). Further studies on *Rana pipiens* racial hybrids. *Amer. Nat.*, **84**, 247–54.
194. MOORE, J. A. (1964). Diploid and haploid interracial hybrids in *Rana pipiens*. *Proc. XI Int. Congr. Genet.*, **2**, 431–36.
195. MOOS, J. R. (1955). Comparative physiology of some chromosomal types in *Drosophila pseudoobscura*. *Evolution*, **9**, 141–51.
196. MOREAU, R. (1966). *The Bird Faunas of Africa and its Islands*. Academic Press, London.
197. MORGAN, P. (1975). Selection acting directly on an enzyme polymorphism. *Heredity*, **34**, 124–7.
198. MORELL, G. M. and TURNER, J. R. G. (1970). Experiments on mimicry. I. The response of wild birds to artificial prey. *Behaviour*, **36**, 116–30.
199. MULLER, F. (1879). Notes on Brazilian entomology. *Trans. Ent. Soc. Lond.*, **3**, 211–23.
200. MULLER, H. J. (1958). Evolution by mutation. *Bull. Amer. Math. Soc.*, **64**, 137–60.
201. MURRAY, J. J. (1972). *Genetic Diversity and Natural Selection*. Oliver and Boyd, Edinburgh.
202. MURRAY, J. (1975). *The Genetics of the Mollusca*. In *Handbook of Genetics*, volume 3, ed. KING, R. C.
203. NAIR, P. S., BRNCIC, D. and KOJIMA, K. (1971). Isozyme variations and evolutionary relationships in the *mesophragmatica* species group of *Drosophila*. Studies in Genetics. VI. *University of Texas Publications*, **7103**, 17–28.
204. NEWELL, N. D. (1967). Revolutions in the history of life. *Geol. Soc. Amer.*, **89**, 63–91.
205. NICHOLS, D. (1959). Changes in the chalk heart-urchin *Micraster* interpreted in relation to living forms. *Phil. Trans. Roy. Soc. Lond. B*, **242**, 347–437.
206. O'DONALD, P. (1969). 'Haldane's Dilemma' and the Rate of Natural Selection. *Nature, Lond.*, **221**, 815–16.
207. O'DONALD, P. and PILECKI, C. (1970). Polymorphic mimicry and natural selection. *Evolution*, **24**, 395–401.
208. OHTA, T. and KIMURA, M. (1973). A model of mutation appropriate to estimate the number of electrophoretically detectable alleles in a finite population. *Genet. Res.*, **22**, 201–4.
209. PARKIN. D. T. (1971). Visual selection in the Land Snail *Arianta arbustorum*. *Heredity*, **26**, 35–47.
210. PERUTZ, M. F. and MITCHISON, J. M. (1950). State of haemoglobin in

sickle-cell anaemia. *Nature, Lond.*, **166**, 677–9.
211. PETIT, C. (1951). Le rôle de l'isolement sexuel dans l'évolution des populations de *Drosophila melanogaster*. *Bull. biol. Fr. Belg.*, **85**, 392–418.
212. PETIT, C. (1954). L'isolement sexuel chez *Drosophila melanogaster*. Étude du mutant *white* et de son allélomorphe sauvage. *Bull. biol. Fr. Belg.*, **88**, 435–43.
213. PETIT, C. (1958). Le determinisme génétique et psycho-physiologique de la compétition sexuelle chez *Drosophila melanogaster*. *Bull. biol. Fr. Belg.*, **92**, 248–329.
214. PETIT, C. (1968). Le rôle des valeurs sélectives variables dans le maintien du polymorphisme. *Bull. Soc. zool. Fr.*, **93**, 187–208.
215. PIPKIN, S. B., RHODES, C. and WILLIAMS, N. (1973). Relation of temperature to the frequency of the *Drosophila* alcohol dehydrogenase allele AdhII. *Genetics*, **74**, S213.
216. POPHAM, E. J. (1941). The variation of the colour of certain species of *Arctocorisa* (Hemiptera, Corixidae) and its significance. *Proc. zool. Soc. Lond. A*, **111**, 135–72.
217. POPHAM, E. J. (1942). Further experimental studies on the selective action of predators. *Proc. zool. Soc. Lond. A*, **112**, 105–17.
218. POST, D. D. and PETTUS, D. (1967). Sympatry of two members of the *Rana pipiens* complex in Colorado. *Herpetologica*, **23**, 323.
219. POWELL, J. R. (1971). Genetic polymorphisms in varied environments. *Science, N.Y.*, **174**, 1035–6.
220. PRAKASH, S. (1972). Origin of reproductive isolation in the absence of apparent genic differentiation in a geographic isolate of *Drosophila pseudoobscura*. *Genetics*, **72**, 143–55.
221. PRAKASH, S., LEWONTIN, R. C. and HUBBY, J. L. (1969). A molecular approach to the problem of genic heterozygosity in natural populations. IV. Patterns of genic variation in central, marginal and isolated populations of *Drosophila pseudoobscura*. *Genetics*, **61**, 841–58.
222. RASMUSON, B., NILSON, L. R., RASMUSON, M. and ZEPPEZAUER, E. (1966). Effects of heterozygosity on alcohol dehydrogenase (ADH) activity in *Drosophila melanogaster*. *Hereditas*, **56**, 313–16.
223. REED, S. C. and REED, E. W. (1950). Natural selection in laboratory populations of *Drosophila*. II. Competition between a white-eye gene and its wild type allele. *Evolution*, **4**, 34–42.
224. REIGHARD, J. (1908). An experimental field study of warning colouration in coral-reef fishes. *Publs. Carnegie Instn.*, **No. 103**, 257–325.
225. RENSCH, B. (1959). *Evolution above the Species Level*. Methuen, London.
226. RICHMOND, R. C. (1970). Non-Darwinian evolution: A critique. *Nature, Lond.*, **225**, 1025–8.
227. RICKLEFS, R. E. (1973). *Ecology*. Nelson, London.
228. RIUBAL, R. (1955). A study of altitudinal races in *Rana pipiens*. *Evolution*, **9**, 322–38.
229. RIUBAL, R. (1957). An altitudinal and latitudinal cline in *Rana pipiens*. *Copeia*, **3**, 212–21.
230. ROBERTSON, F. W. (1955). Selection response and the properties of genetic variation. *Cold Spr. Harb. Symp. Quant. Biol.*, **20**, 166–77.
231. ROWE, A. W. (1899). An analysis of the genus *Micraster*, as determined by rigid zonal collecting from the zone of *Rhynchonella cuvieri* to that of

Micraster coranguinium. *Quant. J. Geol. Soc. Lond.*, **55**, 494–547.
232. SANGER, F. (1956). The structure of insulin. In *Currents in Biochemical Research*, ed. GREEN, D. E., Interscience, New York.
233. SCHARLOO, W. (1971). Reproductive isolation by disruptive selection: did it occur? *Amer. Nat.*, **105**, 83–6.
234. SELANDER, R. K. and KAUFMAN, D. W. (1973). Self-fertilisation and genetic population structure in a colonising land snail. *Proc. Nat. Acad. Sci. U.S.A.*, **70**, 1186–90.
235. SELANDER, R. K. and YANG, S. H. (1969). Protein polymorphism and genic heterozygosity in a wild population of the house mouse (*Mus musculus*). *Genetics*, **63**, 653–67.
236. SHEPPARD, P. M. (1951a). A quantitative study of two populations of the moth *Panaxia dominula* (L.). *Heredity*, **5**, 125–34.
237. SHEPPARD, P. M. (1951b). Fluctuations in the selective value of certain phenotypes in the polymorphic land snail *Cepaea nemoralis* (L.). *Heredity*, **5**, 125–34.
238. SHEPPARD, P. M. (1952). A note on non-random mating in the moth *Panaxia dominula*. *Heredity*, **5**, 349–78.
239. SHEPPARD, P. M. (1967). *Natural Selection and Heredity*. 2nd Edition. Hutchinson, London.
240. SHEPPARD, P. M. and COOK, L. M. (1962). The manifold effects of the *medionigra* gene of the moth *Panaxia dominula* and the maintenance of a polymorphism. *Heredity*, **17**, 415–26.
241. SHEPPARD, P. M. and FORD, E. B. (1966). Natural selection and the evolution of dominance. *Heredity*, **21**, 139–47.
242. SIMPSON, G. G. (1951). *Horses: the story of the horse family in the modern world and through sixty million years of history*. Oxford University Press, New York.
243. SMITH, J. M. (1966). Sympatric speciation. *Amer. Nat.*, **100**, 637–50.
244. SMITH, J. M. (1968). 'Haldane's Dilemma' and the rate of evolution. *Nature, Lond.*, **219**, 1114–6.
245. SNEATH, P. H. A. (1966). Relations between chemical structure and biological activity in peptides. *J. Theor. Biol.*, **12**, 157–95.
246. SOANE, I. D. and CLARKE, B. C. (1973). Evidence for apostatic selection by predators using olfactory cues. *Nature, Lond.*, **241**, 62–4.
247. SOKAL, R. R. and KARTEN, I. (1964). Competition among genotypes in *Tribolium castaneum* at varying densities and gene frequencies (the *black* locus). *Genetics*, **49**, 195–211.
248. SOKAL, R. R. and SNEATH, P. H. A. (1963). *Principles of Numerical Taxonomy*. Freeman, San Francisco and London.
249. SPIESS, E. B. (1968). Low frequency advantage in mating of *Drosophila pseudoobscura* karyotypes. *Amer. Nat.*, **102**, 363–79.
250. STEBBINS, G. L. (1971). *Chromosomal Evolution in Higher Plants*. Edward Arnold, London.
251. STERN, C. (1960). *Principles of Human Genetics*. 2nd Edition. Freeman, London.
252. SVED, J. A., REED, T. E. and BODMER, W. F. (1967). The number of balanced polymorphisms that can be maintained in a natural population. *Genetics*, **55**, 469–81.
253. THODAY, J. M. (1972). Disruptive selection. *Proc. Roy. Soc. Lond. B*, **182**, 109–43.
254. THODAY, J. M. and GIBSON, J. B. (1962). Isolation by disruptive selection.

Nature, Lond., **193**, 1164–6.

255. TINBERGEN, L. (1960). The natural control of insects in pine woods. I. Factors affecting the intensity of predation by song-birds. *Archs. neerl. Zool.*, **13**, 265–336.
256. TSCHERMAK, E. VON (1900). Über künstliche Kreuzung bei *Pisum sativum*. *Ber. dt. bot. Ges.*, **18**, 232–9.
257. TURNER, J. R. G. (1971). Studies of Müllerian mimicry and its evolution in burnet moths and heliconid butterflies. In *Ecological Genetics and Evolution*, ed. CREED, E. R., pp. 224–60, Blackwell, Oxford.
258. VETUKHIV, M. (1954). Integration of the genotype in local populations of three species of *Drosophila*. *Evolution*, **8**, 241–51.
259. VETUKHIV, M. (1956). Fecundity of hybrids between geographic populations of *D. pseudoobscura*. *Evolution*, **10**, 139–46.
260. VETUKHIV, M. (1957). Longevity of hybrids between geographic populations of *D. pseudoobscura*. *Evolution*, **11**, 348–60.
261. VIGUE, C. L. and JOHNSON, F. M. (1973). Isozyme variability in species of the genus *Drosophila*. VI. Frequency-property-environment relationships of allelic alcohol dehydrogenases in *D. melanogaster*. *Biochem. Genet.*, **9**, 213–27.
262. VOLPE, E. P. (1954). Hybrid inviability between *Rana pipiens* from Wisconsin and Mexico. *Tulane Stud. Zool.*, **1**, 111–23.
263. VOLPE, E. P. (1957). Embryonic temperature adaptations in highland *Rana pipiens*. *Amer. Nat.*, **91**, 303–10.
264. WALLACE, A. R. (1889). *Darwinism : an exposition of the theory of natural selection*. Macmillan, London.
265. WALLACE, B. (1955). Inter-population hybrids in *D. melanogaster*. *Evolution*, **9**, 302–16.
266. WALLACE, B. (1958). The role of heterozygosity in *Drosophila* populations. *Proc. 10th Int. Congr. Genetics*, **1**, 408–19.
267. WALLACE, B. (1968). *Topics in Population Genetics*. W. W. Norton, New York.
268. WATSON, H. C. and KENDREW, J. C. (1961). Comparison between the amino acid sequences of sperm whale myoglobin and of human haemoglobin. *Nature, Lond.*, **190**, 670.
269. WATT, W. B. (1968). Adaptive Significance of Pigment Polymorphisms in *Colias* butterflies. I. Variation of Melanin Pigment in Relation to Thermoregulation. *Evolution*, **22**, 437–58.
270. WELDON, W. F. R. (1901). A first study of natural selection in *Clausilia laminata*. *Biometrika*, **1**, 109–24.
271. WILLIAMSON, M. H. (1960). On the polymorphism of the moth *Panaxia dominula*. *Heredity*, **15**, 139–51.
272. WOODWORTH, C. M., LENG, E. R. and JUGENHEIMER, R. W. (1952). Fifty generations of selection for protein and oil in corn. *Agron. J.*, **44**, 60–6.
273. WRIGHT, S. (1922). The effects of inbreeding and crossbreeding on guinea pigs. *Bull. U.S. Dept. Agric.*, **1121**, 1–59.
274. WRIGHT, S. (1929a). Fisher's theory of dominance. *Amer. Nat.*, **63**, 274–79.
275 WRIGHT, S. (1929b). The evolution of dominance: comment on Dr. Fisher's reply. *Amer. Nat.*, **63**, 556–61.
276. WRIGHT, S. (1932). The role of mutation, inbreeding, crossbreeding and selection in evolution. *Proc. 6th Int. Congr. Genet.*, **1**, 356–66.
277. WRIGHT, S. (1933). Inbreeding and homozygosis. *Proc. Nat. Acad. Sci.*

U.S.A., **19**, 411–20.
278. WRIGHT, S. (1943). An analysis of local variability of flower color in *Linanthus parryae*. *Genetics*, **28**, 139–56.
279. WRIGHT, S. (1948). On the roles of directed and random changes in gene frequency in the genetics of populations. *Evolution*, **2**, 279–94.
280. WRIGHT, S. (1965). Factor interaction and linkage in evolution. *Proc. Roy. Soc. Lond. B*, **162**, 80–104.
281. YAMAZAKI, T. (1971). Measurement of fitness at the esterase-5 locus of *Drosophila pseudoobscura*. *Genetics*, **67**, 579–603.
282. YARBROUGH, K. M. and KOJIMA, K. (1967). The mode of selection at the polymorphic Esterase-6 locus in cage populations of *Drosophila melanogaster*. *Genetics*, **57**, 677–86.
283. ZUCKERKANDL, E. and PAULING, L. (1965). Evolutionary divergence and convergence in proteins. In *Evolving Genes and Proteins*, eds. BRYSON, V. and VOGEL, H. J., pp. 97–166. Academic Press, New York.

Index

Bradford College Library
QH430 .P367
BCHA
Parki